Mastering Project Management

Surviving the Maelstrom and Delivering on Commitments

R. J. Monson, PhD

For Mary – my sweetheart

and

To those who seek to make the world a more complicated place, and who choose not to take their promises seriously.
This book is not for you.

Table of Contents

On the Cover:

A surfer rides the maelstrom that is created where the dynamic ocean surf meets the constraining land. This is where the chaos and turbulence is most profound, and the surfer has learned to utilize the inherent energy and dynamics to his advantage, precariously operating at the edge of chaos. This is similar to how the dynamic work to be completed in the project meets the constraints of the organization, resulting in turbulence and chaos. This is the domain of project management, and the effective project manager must learn how to harness the inherent energy created in this environment to achieve his objectives.

Introduction

This text is intended as a basic instructional approach designed around a single semester Project Management curriculum. The design of the course introduces the student to many of the basic concepts of Project Management, culminating in a wide variety of examples provided to demonstrate potential applications of the methodology and reduce the anxiety of the first time practitioner. In general, Project Management is not a technically challenging domain; it is more a discipline that requires fervent attention to detail in order to achieve success.

Design of the course is systematic and aligned with the typical execution of a graduate program. This course and text have been designed to create a "how-to" methodology that will enable students to recreate project management in their own organizations.

The focus of instruction is on walking the student through the traditional Project Life Cycle. Beginning with the identification of stakeholders and users, and progressing through the methodology and final acceptance of the project, the student gets a taste of the myriad skills needed to successfully execute a complex system.

The course relies upon tools and techniques developed within many organizations including commercial corporations, non-profits and the United States Department of Defense, as well as upon academic studies undertaken to further the project management domain. Although this methodology is capable of supporting very large programs, the concepts are easily applied to smaller efforts typically facing medium and small businesses.

Thus, this Project Management text, Mastering Project Management, is a pragmatic and practical method to develop project management capabilities for the inexperienced student as well as for the seasoned practitioner.

Why do we need Project Management?

As we advance a variety of technologies forward and innovations become commonplace in electronics and software, we discover a rapid explosion in the complexity of work. No longer is it usual for us to enter a machine shop and start to assemble parts to create a product that will be offered in the marketplace. We can no longer sit down and begin coding a piece of software in the hopes that as we add more complexity of detail, we will create a valuable application that people will be willing to purchase. We are instead discovering that work's complexity requires us to devote appropriate time considering the best approach and method for ultimately producing a product worthy of the marketplace.

If we liken a typical work effort in our organization to a piece of art such as a painting, we can begin to see this complexity increase. Imagine your art class in elementary school, when the teacher suggested that you undertake an artwork in the form of a painting. Most likely, you simply started with a blank canvas and began adding paint without a dramatically detailed plan in mind. Imagine now that you are a professional painter, where you are paid significant amounts of money for your work. In this case, when someone asks you to undertake an artwork, you likely have neither the time nor the inclination to waste any effort with regard to the challenge. In fact, you would likely paint this artwork as efficiently as possible, minimizing the input of your effort from the start to the finish of the work.

In the same way, we have undertaken challenging work within our workplace, for which we are compensated. It is in both our best interest as well as our organization's best interest that we accomplish this work as efficiently and capably as possible. Therefore, it seems clear that if methods have been developed that enable us to produce more effective results with less input, it is in our best interest to take advantage of those methods and processes in order to act as good stewards for our organizations.

In addition to the increase of complexity in work, we must also account for certain human limitations. These are related to a variety of human factors that result in some of the challenges we will all face in the project-management domain. These include issues of communications, sociology and psychology, all of which may contribute to misunderstandings or misapprehensions within the workplace.

As an example, people have a natural tendency to undertake tasks that are easy rather than those that are hard. People will also likely choose to do tasks that are short as opposed to those that take a long period of time. Humans have a tendency to undertake tasks that fit into the time remaining, as opposed to those that might extend beyond the time readily at hand.

When one undertakes a challenging and complex project, we

understand that much of the work will be difficult and will take a considerable amount of time. Likewise, those stakeholders that are interested in the progress of our project will also be required to do tasks that are difficult and could be time-consuming. These tasks might include taking the time to consider the requirements or the scope of the work to be completed. If, however, the individuals choose not to put in their time to address these issues, often they will have misconceptions or make dramatic assumptions regarding the scope of the work to be completed. Due to these limitations, it is incumbent on the project manager to ensure that all stakeholders in the project are satisfied as well as satisfactorily apprised of the work being completed throughout the duration of the project development. Therefore, if the project manager is unaware that stakeholders will likely not spend considerable time in reflection regarding the project, the project manager will face additional effort in managing these individuals. There are many other issues related to human limitations that will directly and indirectly impact the project manager. It is therefore important that the project manager take advantage of systems that have been developed in order to address these issues.

Project managers are frequently required to justify that their approaches are both beneficial to the organization as well as fiscally responsible. This implies that the project managers are producing a maximum output with a minimal input to the project in the form of resources. Therefore, project managers are fundamentally interested in providing a coherence of effort amongst the disparate members of their project teams. To enable this coherent effort, they must establish a common understanding among the stakeholders in order to assure that all involved are on the same page.

This can be particularly frustrating for a typical analytical project manager, who has a fundamental understanding of what needs to be accomplished on the project and assumes all others have reached identical conclusions. Without spending sufficient time assuring that a common understanding exists, it is unlikely the project manager will enjoy a coherent effort amongst a diverse team of project members. Therefore, it is of practical importance to utilize the methods developed to address these challenges in the practice of project management.

Finally, it is important for project managers to recognize that they are fundamentally required to predict the future. This is largely due to the fact that they are expected to produce a project development plan that will take into account all of the various moving parts of the organization as well as the project development itself, and produce appropriate deliverables that will satisfy all of the project's stakeholders. As we might expect, many of the factors associated with projects will change throughout the duration of project development. The world will change, the people will change, the company will change, and more than likely the objectives of the project will change, as well. Amidst all of these changes, project managers are expected to keep their projects on appropriate paths that will result in project success.

Many project management tools have been developed to help the project manager to predict the future outcome of the project. In general, the ability to accurately predict the technical performance, delivery date, and expected cost is valued because it leads to more efficient efforts and less turbulence within our organizations. Thus, the ability to accurately plan and predict is essential for every successful project manager.

What does Project Management accomplish?

Many diverse practitioners in the project management domain have devoted long hours in the development of the project management methodology that is shared in textbooks and courses around the world. Through the development of technologies and products for the marketplace, thee contributors have defined an effective process to aid their colleagues and organizations in effective project management. An agglomeration of these efforts has been painstakingly collected to build a common body of knowledge that exists in many forms and utilizes many different methods, dependent upon the source of information. However, they share common ground and are often reflections of one another in their documentation.

While a given project management practitioner may select one or another of these services as his or her preferred terminology, semantics and method, no single source contains a magic formula that eases all potential challenges. In fact, most project managers rely upon multiple sources in order to gain a large toolset of capabilities that can be utilized dependent upon the situation.

The project management methodologies largely focus on methods that will allow the project manager to:

- Clarify intentions and expectations for the project
- Define the approach for communication to the stakeholders

- Establish alignment of the development team and stakeholders
- Develop aggregated assignments that may be used to piece out the work, enabling synergy in the project team

Whatever the source, it is also good practice to pay close attention to the project managers you have the opportunity to observe, learning from their successes and failures, as well as from your own. Project management is a highly personal and individual endeavor, and what works for one individual may or may not work for another. All that we can be sure of is that if we choose to remain in one place for too long, we will likely be left behind. Project management is not just a skill to be utilized in the workplace, it is a lifelong learning experience that will enable a person to accomplish more and improve his or her satisfaction with the results. It is a fascinating journey that awaits each of us.

What are the typical challenges faced by Project managers?

Many studies have been conducted regarding the potential success or failure modes of engineering projects in the hope of identifying the predictors of a project's failure or success. These predictors might be very useful in the development of appropriate project management methodologies and approaches for specific practitioners.

Summaries are provided for many of these studies in order to define the common factors typically described as predictors of success or failure in project management endeavors. Further information on the studies may be obtained by examining the referenced articles in the bibliography.

Key Elements of Project Outcomes (Standish Group CHAOS results, 1996)

Top Success Factors

- User involvement
- Executive Management support
- Clear requirements
- Proper planning
- Realistic expectations
- Smaller project milestones
- Competent staff
- Ownership
- Clear vision and objectives
- Hard working, focused staff

Top Failure Factors

- Lack of user input
- Incomplete requirements
- Changing requirements
- Lack of executive support
- Technology incompetence
- Lack of resources
- Unrealistic expectations
- Unclear objectives
- Unrealistic timeframes
- New technology

Critical Success / Failure Factors in Projects (Belassi & Tukel, 1996)

- Top management support throughout
- Proper goal definition and setting
- Accurate estimates
- End-user involvement
- Project manager on site
- Appropriate team selected
- Testing and training

Patterns of Software Systems Failure and Success (Jones 1996)

Project Management Failure Factors

- No historical software measurement data
- Failure to use automated estimation tools
- Failure to use automated planning tools
- Failure to monitor progress of milestones
- Failure to use effective architecture
- Failure to use effective development methods
- Failure to include formal risk management
- Failure to use formal configuration management

- More than thirty percent creep in requirements

Principal reasons for DOD software project failures, Report of the Defense Science Board Task Force on Acquiring Defense Software Commercially (1994)

- Poor requirements definition
- Inadequate software process management
- Lack of integrated product teams
- Ineffective subcontract management
- Too little attention to software architecture
- Poorly defined, inadequately controlled interfaces
- Limited or no tailoring of military standards

Software Systems Failure Modes (Website: http://suraj.lums.edu.pk)

Hostile culture

Punishing bad news messengers

Poor reporting structure

Sponsors are not involved

The Success of any project is linked to the people

Take care of the people to succeed

Over-commitment

Unclear expectations and understanding of the scope

Decision Making Escalation

Movement becomes more important than progress

Political Pressures

Hidden Agendas

Technology-focused developments

Ignoring the Human

The lure of the leading edge

It can be costly and heartbreaking

Complexity underestimated

Simplified by individuals who are not informed
Inadequate consultations with major stakeholders
User needs in the dark
Design by committee
No major responsibility holder
Technical fix for a management problem
Lack of whole picture thinking
Staff turnover
Lack of engagement
Competency
Decisions made by the wrong people
Receding deadlines
Soft plans with malleable objectives and milestones
Inadequate testing and training
Not having discipline in our development
Communication problems and work group problems
As always

Most of these observations will not surprise an experienced project manager. However, it is not adequate to identify the issues as much as it is crucial to identify solutions to the existing problems. To do this, we will consider not only the factors that indicate or predict project failure, but also the mechanisms that are activated when these scenarios present themselves. With this information in hand, we should be able to better understand how to circumvent these issues, and additionally, we can use this knowledge to define the specific areas of importance we should address in our project management methodology.

Therefore, we will examine the following common factors as indicators of potential project failures. Each of the identified predictors for project failure will also be described with regard to the typical failure mode created for the project.

From this information, we hope to identify areas of importance in our project management methodology in order to establish a robust and effective project management approach in this text.

15 Common factors predicting project failure:

1. Lack of user input

   At the onset of a project the stakeholder or user is asked about the project's needs. The response is often considered naïve or ignorant, and from that point forward the user is largely ignored during the development cycle. This may be manifest in a development team's perspective that the user doesn't really understand the problem or that the user doesn't truly understand what is needed.

2. Vague or Incomplete requirements

   Program managers frequently provide vague or incomplete project requirements that are written in a nebulous format or only refer to the provider's details of interest. This may result in the belief that as long as the stated objectives are achieved, the project will be deemed a success. In reality, the team is responsible for the development of a complete and detailed requirements list.

3. Changing requirements / Scope Creep

   Many projects exhibit significant scope creep

during the development phase. While it seems reasonable to expect the requirements to remain static during the development phase, the system is likely to undergo spontaneous and regular change throughout the cycle due to the changing insights and understanding associated with the development process.

4. Lack of leadership support

   Many projects are assigned an executive or a member of the organization's leadership team to shepherd the project through the development cycle. This is usually in addition to executive's regular job duties. For this reason, if he or she is not intimately interested in the success or failure of the project, the project team may have unrealistic expectations regarding his or her contributions to the development effort.

5. Technology challenges

   Project development efforts often seek to make dramatic technological changes. Often, these technical challenges are blamed for lack of progress or lack of completion of the development process. However, virtually every instance of project management introduces new technologies, processes or methods. In the event the development team is not managing the changes, they may find that the perceived success of their efforts is less than desired.

6. Lack of resources

In any organization there are limited resources available for development work. This is often caused by the fact that many projects within the development cycle will overrun their initial schedule and budgets in their team's attempts to provide the desired outcomes. By virtue of this, if the leadership of the organization has planned their project efforts at one hundred percent resource utilization, in the midst of a development cycle it will be discovered that there are not enough resources to go around. This will often result in confidently managed projects losing their resources to those that are managed less competently. In short, high performers are punished.

7. Unrealistic expectations

Project managers often expect the customer to provide detailed requirements and expectations. Unfortunately, it is unrealistic to expect that a customer is able to describe the expected outcome of the project with sufficient detail to allow the project manager to proceed without investing significant effort in requirements development. A successful project manager must therefore ensure that the customer's expectations are appropriate and in line with the project's planned deliverables.

8. Unclear objectives

In much the same way, the project manager may allow project stakeholders to assume the project objectives that are intended, or expect that those people peripheral to the project team have a clear understanding of the objectives being pursued by the project team. Unfortunately, when multiple individuals are asked to define and record objectives for a complex project development effort, the objectives they define may lack coherency and consistency.

9. Unrealistic timeframes

   Similarly, if the project manager is not clear about the expectations for project schedules and expected completion dates, the stakeholders will often assume dates of convenience that will become expectations within the project.

10. No historical project management data

   Project managers are expected to create detailed plans describing the work to be accomplished during the development phase. This implies that the project manager is expected to predict the future within the constraints of the development effort. As might be expected, project managers will exhibit both good and poor estimation abilities, and in the areas where they have less expertise will often rely upon historical artifacts in their estimation efforts. If these artifacts are not available, and if project managers do not have enough experience to rely upon their own estimates, it is likely that the estimates provided will be unrepresentative of the project's outcome.

11. Did not monitor progress of milestones

    While it is important for project managers to make long-range plans regarding their development efforts, plans may be revised and clarified as information becomes available during the execution phase of the project. Therefore, it is likely that the fidelity of the project planning will be higher in the short-term and lower in the long-term. If the project manager is providing regular progress updates and paying close attention to the predicted and actual outcomes, this information may allow appropriate revisions to the detailed project plans to better predict the outcome of the project.

12. Lack of project management methods

    Recommended project management methodologies have been developed over a long period of time by people who have experienced challenges and difficulties in the execution of their project management duties. Most project managers will run into challenges and difficulties similar to those experienced by those who have gone before them. Therefore, it is of significant benefit to be familiar with those techniques recommended by experienced project managers in order to shorten the personal development necessary to become an effective project manager.

13. Lack of risk management

During the development phase, significant attention must be paid to the items that present a risk to the successful completion of the project. Keeping in mind that the original project plan is an attempt to predict the future, the project manager must realize throughout the process that uncertainties often equate to changes in both schedule and budget. By virtue of this, risk management is an essential skill that will enable more realistic prediction of the project outcomes.

14. Lack of engagement on the team

Even though the project manager has produced a highly representative development plan and has presented that to the project team, the project is not guaranteed any measure of success. In fact, if the team members are disengaged or disinterested, it is unlikely that project deliverables will be finished on time and budget. Careful consideration must therefore be given to the engagement of the team and the proposed project plan. Without appropriate commitment to completion, it is unlikely the planned outcomes will be realized.

15. Poor integration

When projects cross-functional boundaries such as between mechanical and electrical systems or between business and manufacturing systems, the interfaces and the integration across these boundaries must be addressed. Lack of effort in this area will often result in a developed system that is made up of largely functional components that do not achieve the desired outcome of the project development effort. In short, you may end up with a mechanical system that performs well and a software system that does not allow appropriate control or interfacing, or a business system that does not support effective manufacturing of the end product. In both cases, a lack of integration has resulted in a less efficacious project outcome.

How does Project Management fix these issues?

When we consider each of the predictors previously discussed, we realize that it is important to understand how our proposed project management methodology will address these challenges and result in a more successful project management effort. We will consider each of these cases in groups and reflect on how they are addressed by appropriate project management methodologies.

Lack of user input / Vague or Incomplete requirements / Changing requirements / Scope Creep

A successful project manager must take sufficient time to explicitly clarify a project's objectives and deliverables at its outset in order to ensure its successful execution. This requires the project manager to spend sufficient time with the project's stakeholders to define these parameters and ensure that the planned project will meet the stakeholders' needs. This also requires the project manager to design the project to allow effective prediction of its outcomes. In short, it is not enough to define a project plan that can be executed effectively; it is also imperative to ensure the outcomes of the project will satisfy the stakeholders' needs and desires.

Lack of leadership support / Lack of engagement on the team

Project management utilizes an organization's resources to achieve desired project outcomes. This requires the project manager to go beyond the assignment of resources and secure the organization's leaders' interest and investment in the outcome of the project. The project manager must also convince the assigned members of the project team to make significant efforts towards a common goal in a manner that will result in the desired project outcomes.

If the project manager fails in either of these areas, it is likely that the project will not be a successful one. As an example, if the leadership support is lacking within the organization and resources become tight, it is likely those resources will be removed from those projects less interesting to the organizational leadership team. In the event the project manager does not make significant effort to engage the project team, it is likely that some members of the project team will contribute at high levels while other members contribute at a diminished level. If the project manager expects to achieve a high level of effort with each member of the project team, he or she must engage the team in the outcomes of the projects sufficiently to enhance their motivation to contribute effectively.

Therefore, it is important for the project manager to plan and communicate the project deliverables and outcomes to the project team in a manner that will not only establish the items of interest for the project but also the reasons why the project should be completed as planned. In other words, the project manager must create a compelling story so that the team will choose to contribute a high level of effort towards the successful completion of its outcomes.

This is generally achieved in our project management methodology by means of careful attention to project alignment and the appropriate strategy for the project. If this is accomplished effectively, the project plan and the reason why it is an optimal solution will be communicated to the team, and will greatly enhance their engagement and the probability of a successful outcome.

Technology challenges / Poor integration

Project managers are essentially change managers within organizations. If they are not incorporating change, there is no need for them to seek their projects' outcomes. For this reason, effective project managers must recognize that their ability to manage change effectively will contribute significantly towards the success of their projects. They must also spend time during project development managing the changes being incorporated into the organization, in the hopes that the effective socialization of these changes will enable people to accept them more readily and create a higher probability of success for projects' outcomes.

Likewise, successful project managers who take the time to communicate the intended changes effectively throughout the development team are more likely to enable an understanding that crosses the functional boundaries and limitations of the varying departments within the project team. This focus on effective communication and the breaking down of barriers will result in a more highly integrated and less problematic implementation for the project. Therefore, the project manager must become skilled at managing the communications and information transfer that naturally occurs in any project-management endeavor.

Unrealistic expectations / Unclear objectives / Unrealistic timeframes

Project managers are not only responsible for the outcomes of the project; they are the architects of the change to be implemented. For this reason, they must understand that the definition of objectives and outcomes is entirely at their discretion. Once they have established the expected outcomes for the project, it is also imperative that they are able to communicate these effectively to the various stakeholders on the project. Once this communication has taken place, they will often find that they need to negotiate the potential project outcomes with each of the stakeholders in a manner that will allow them to coherently construct a representative project plan.

Therefore, project managers must enhance their skills in the project planning domain, as well as the ability to make decisions regarding the outcomes to be delivered by the project. Additionally, they must become effective negotiators in order to address those conflicts and limitations that will naturally occur amongst the stakeholders on any given project effort.

Effectively planning proactively and then communicating this expected plan to the stakeholders would often tend to alleviate any unrealistic expectations or assumptions that have been made by those stakeholders. In fact, most of these unrealistic expectations are created in the absence of and effectively communicated project plan.

No historical project management data / Lack of resources

A major challenge facing most organizations practicing project management is the limitation of resources for application on those projects. This is typically caused by a lack of understanding of what resources will be required to complete this project effort. Often this lack of understanding is brought about by a lack of reflection of past work that has been accomplished within the organization, although it may also be caused by a lack of experience within the organization.

As might be expected, a new project manager will not be able to effectively estimate the duration or cost of their project as they have never accomplished this successfully in the past. One reason why a postmortem is usually recommended on most projects is to provide feedback to the original project planner regarding their performance as an estimator on the project. If no historical project management data is present in the system or has in the past been used to temper the expectations of individuals within the organization, it is unlikely that people will have a good foundational understanding of what it takes to complete a complex project effort.

When this is the case, it is not uncommon for organizations to plan for a 100% utilization of their resources throughout the performance period. As projects are delayed or budgets are exceeded, the organization will likely discover that they have over-planned their resources and they will subsequently become extremely scarce within the organization. In many cases, those projects that are being effectively run will be robbed in order to make up for the sins of other project managers who are not keeping their efforts on plan.

To combat these challenges in our organizations, we must more readily utilize past project performance data in order to better predict future project performance data. This is important on an individual basis as well as on a historical basis for our organization. If, for example, you know that in the past you have typically overrun all software projects by 200%, this value might be an extremely useful predictor for future software project performance estimates.

When we become effective at estimating, and we become more accurate in our prediction of the duration and cost of the project, we can then more readily plan for the use of 100% of our resources. If we are in a position where we cannot effectively plan to use of our resources in a predictive fashion, then the cycle will continue to repeat itself and we will ultimately find ourselves canceling healthy projects in order to provide resources to those that are less healthy. This is obviously a sub optimal solution for our organizations, thereby establishing the strategic significance of developing effective project management practices.

Lack of project management methods / Lack of risk management / Did not monitor progress of milestones

As has been previously argued, the lack of effective project management practices can be extremely costly for our organizations. It is of supreme importance to become effective at predicting the duration and costs of work that must be accomplished by our organization, in order to allow for appropriate planning and prediction within the constraints available to us.

Due to this need to effectively project the magnitude of work to be accomplished, we must also become adept at assessing the risks associated with any project effort. These consist of unpredictable project components that impact the schedule, budget or technical performance, and ultimately adversely affect the project and adjacent projects within the organization. Risk management is a methodology that has been developed in order to ascribe a magnitude to the pending risk on any given project, thereby allowing more predictability with regard to the expected project outcomes.

Due to the complexities of changing environments and project elements, most project managers exist in a highly dynamic environment. Because of this, that they must enhance and update their project plans on a regular basis. These updates typically reflect to knowledge that has been gained on the project, or insights that have been developed due to the occurrences on the project. By virtue of this, short-term planning is of a much higher resolution than longer-term planning. When resolution is added to shorter-term plans, these may reflect positively or negatively on the longer-term plans, thereby allowing us to extrapolate the current results to provide better prediction regarding future expectations on our projects. To accomplish this, we typically monitor the status of our projects with regard to both the cost performance as well as the schedule performance. This assessment enables us to make representative predictions of the likely outcome of our project during the execution phase. This will in turn allow us to inject less turmoil into the organization due to schedule or budget performance issues brought about by our project not meeting the original plan.

The Project Life Cycle

Project Management abilities are the least effectively taught and most critical skills that successful engineers must possess. These skills transcend the usual engineering or business management abilities taught to leaders, as they become life skills that can be used on a day-to-day basis and are useful outside the Engineering domain. They have become critical because research suggests virtually all white-collar work is project-based. In addition to this work focus, most technical professionals also seek to plan their personal lives around a project orientation. Therefore, it is imperative that any effective leader in an Engineering organization demonstrates appropriate and pragmatic Project Management skills.

Many taxonomies exist that describe the project life cycle; the Project Management Institute (PMI) adopts one of the most common. This life cycle incorporates a five-phase approach to project understanding, shown in Figure 1.

Figure 1: The Project Life cycle

The first phase of a project is Project Initiation. This is focused on shaping the project in a manner that will be appropriate for the individual, the project team, and the organization for which the project is being completed. This implies that the project leader is well acquainted with the organizational goals and strategy and potentially many of the tactics being implemented within the organization. With this knowledge, the practitioner can plan an appropriate project, which will fit well with the expectations and needs of the organization. This alignment is critical to ensure the effective application of resources within the organization, and often a minimum investment of time is applied to this phase. This can be problematic in later phases, as the project being accomplished may be less suitable than other projects for the future success of the organization.

The second phase of the project life cycle is Project Planning. This is the heart and soul of most project managers' day-to-day existence. It involves defining the specification, schedule and budget for the project deliverables, in order to lay out appropriate resource needs, timelines and funding that are required to complete the project as specified. If a project manager was only to become well versed in a single phase of Project Management, Project Planning is the most suitable. With that being said, Project Planning is typically not adequate to expect project success, and the Project manager must continue to further develop skills until a level of mastery in all phases of the project life cycle is achieved.

The third and fourth phases of the project life cycle are typically concurrent in any given development environment. Phase three focuses on Cost Control, which is the establishment of appropriate metrics and status methods for the project. It is required to maintain updates and communications for other members of the organization, and to accurately project the likely outcome of the project at any given time. Effective Cost Control is highly useful and predictive, and it enables the Project manager to take action quickly in the event of a downturn in project status. Additionally, utilizing a well-established methodology such as Earned Value allows easy communication across project boundaries due to a consistent and practical status system.

The fourth phase of the project life cycle is Project Execution. This focuses on the ability of the Project manager to engage the workforce and to enable that workforce to accomplish the objectives and deliverables set out in the project plan. This involves the application of the soft skills of management, which include teaching, learning, communication, motivation, relationship building and other related skills that are often grouped under the classification of emotional intelligence. While the project-planning phase is the most significant for the future success of the project, an ability to completely execute a project effectively is a skill not easily mastered by many technical professionals. This skill is typically not valued highly by those professionals because it requires a focus on the people as opposed to the technical aspects, something which many project professionals are ill-equipped to handle. Therefore, effective Project Management will require the effective execution of the project in addition to appropriate planning and monitoring.

The last phase of the project life cycle is known as Project Close-out or Closure. This phase is intended to require the project manager to revisit the original project plans and compare them with the actual executed history of the project in order to measure the accuracy of the original projections and estimates. This assessment is expected to improve the future planning and estimating capabilities of the project leaders that will manage multiple projects throughout their career. Often this phase gets ignored when resources become tight, and project leaders move on to their next challenges. This is unfortunate, as project planners are then unable to assess their previous estimation capabilities, thereby limiting their readiness to more effectively plan and execute projects in the future. Effective project managers will ensure that an appropriate project postmortem and Closeout is accomplished in order to achieve these goals.

Phase 1: Project Initiation

Project Initiation is a critical aspect in the development of effective projects within the organization. Engineering leadership should take great care to assure that Project Initiation is accomplished on all projects, as this sets the stage for project alignment and the effective support of programs within the organization.

Defining Projects:

A project is a temporary endeavor undertaken to create a new product or service (Project Management Book of Knowledge – PMBOK.)

Project Management:
Project Management is the application of knowledge, skills, tools and techniques to project activities in order to meet or exceed stakeholder needs and expectations from a project (PMBOK).

Project Management largely consists of Two Parts:

Project Planning

- The Requirements of the work
- The Quantity of the work
- The Resources needed for the work

Project Control and Execution

- Measuring and assessing status
- Making adjustments to the plan
- Managing the engagement of the team

Appropriate initiation of any project requires considerable thought and reflection on the organization for which the work is to be performed. The project leader must consider what strategic imperatives are currently underway within the organization, and which competitive priorities might be most appropriately applied in the execution of the project. Once the project leader has a clear understanding of this alignment, it is appropriate to move forward in the development of the initial project plan.

To do so, a project leader begins at the definition of the goals to be achieved by the project. These will differ from the organizational goals, but should be in alignment with or in support of those business goals. Therefore, to begin our plan we must clearly define our goals in a narrative form.

Define the overall intent of your project – **The Goal**

As an example, if we are responsible for the development of a new product and a chocolate chip cookie company, we must discern the ultimate goal of our project. We might be tempted to suggest that it is to dramatically increase our market share, or increase our profitability. This type of goal can be problematic, for the same reason that the application of a business strategy is challenging for an engineering organization. It is very difficult to know which recipe might be best suited to increase market share, although one might instead change that focus a bit and produce an effective goal statement for the project team.

Goal statement: Provide a new chocolate chip cookie product that will taste better than the competitors' products.

This goal statement is effective because it clearly delineates the criteria used to define the appropriate project plan to accomplish the ultimate business goal of increasing market share. Our next step, then, is to begin to define how this will be accomplished. As the reader will recall, this is synonymous with defining the strategy of the project, and how this strategy aligns with the organizational strategy. This is typically referred to as the project mission and is delineated in the mission statement.

To begin to understand how we might approach this question, we might consider additional viewpoints, such as:

How do you intend to accomplish this work? What is your general approach?

What key skills and abilities will make your approach the correct one that will benefit the organization the most?

Consideration and reflection upon these questions will help to define for the project leader the competencies or skill areas of the resources that can best be applied on the project. Utilizing our example of a chocolate chip cookie company, we know that our previous strategy was to consult the finest chefs for recommendations of recipes and techniques and to use the finest ingredients available in the preparation of our cookies. Since this is the case, and we have been able to effectively apply these skills in the past, it seems appropriate for us to continue to focus on areas where we have demonstrated a significant capability or differentiator over our competition. Please also note that if the intention is to right a sinking ship, it may be appropriate to make a significant departure from past strategies in order to become more competitive or to remain competitive in the marketplace.

In this case we might choose to answer those questions by focusing on a general approach to increase the quality of ingredients that are utilized in our cookies. For example, a portion of our project may include research into what makes a cookie better tasting to the consumer, as well as research into what consumers believe defines a better quality ingredient. Once we have established this understanding we can define our mission statement for the project, which is a sentence or two describing how we will accomplish the goals. Another way to think of this is that the mission of the project should describe simply why your customer wants you to accomplish the project.

Project Mission: Produce an innovative cookie that the average consumer believes taste better than the competitors' cookies.

It is important to understand that this mission statement should reflect one of the five competitive priorities previously defined, including cost, quality, dependability, flexibility, or innovation. In this case we have elected to take an innovative approach to produce the best tasting cookie, which previously did not exist. This might also imply that we could charge a premium for this product.

Next, we will define a statement of work for the effort to be undertaken. This statement of work should represent an objective view of the mission to be accomplished. By objective, we mean that the final statement must be both measurable and achievable in order to establish that we have completed the deliverables we have set out to complete. Some care must be taken not to write this in a goal orientation and to assure that the final result is provable as having been completed.

Goal: A vision or intent that may or may not be achievable

Objective: A measurable and achievable goal

Statement of work: Develop and document a new chocolate chip cookie product utilizing higher milk-fat chocolate and all natural dairy ingredients to produce the best tasting cookie for our customer.

Finally, the purpose of the project should be defined. This should describe why the organization is undertaking the completion of the project.

Purpose: To provide a new chocolate chip cookie product that will result in the development of a new market of high quality cookies with superior flavor for our company.

Competitive Strategy and Product Alignment

In most organizations, the engineering leader has limited resources available to accomplish the work required. As plans go awry and changes pop up within the organization such as resignations, sickness, and maternal or paternal leave, exquisitely detailed schedules become less relevant or even blatantly incorrect. Departmental plans become derailed, and engineers find themselves buried beneath so many demands that it is overwhelming. Many engineering leaders must accept that they have not given much thought to the complexities of multiple project environments, and instead rely upon the heroic efforts of many to keep the department afloat. In addition, many leaders do not communicate their resource planning and scheduling effectively throughout the company, and thereby often place unrealistic demands on their engineering associates. How can we possibly transform this maelstrom of changing demands, priorities and ground rules into a more stable and predictable environment? We can do so by embracing a strategy as a ground truth in the midst of Chaos, and by using the strategy to prioritize and guide our decisions.

Frequently, engineering leaders find they have significant work to be accomplished with less than adequate resources. Therefore, it is very important for leaders to most efficiently use the available resources to accomplish as much as possible. In short, given the scarce resources at hand, how does a leader ensure that they are applied to the most pressing problem or most beneficial opportunity? We utilize strategy to create focus and coherence in our decisions.

In order to completely understand the role of strategy and its value within the engineering context, it is important to consider the appropriate alignment of purpose throughout the different levels within the organization, and why that alignment may enhance or limit the organization's ability to compete effectively. We will begin by considering a simple construct of the organization, comprising three specific levels of responsibility. This may be seen in Figure 2.

The Goal-Strategy-Tactics Tasks Framework

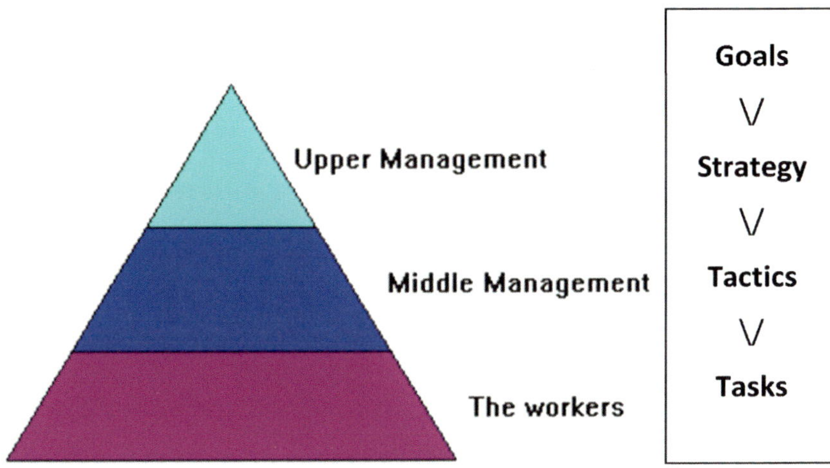

Figure 2: Goals - Strategy - Tactics - Tasks Framework

At the top of the organization, we find the executives of upper management. Upper management's purpose within the organization is supposed to be focused upon the development of an appropriate strategy and vision for the business. This vision or strategy is intended to communicate to the other members of the organization the direction and focus, to be used instead of relying solely upon a centralized decision maker. In other words, the CEO does not need to make all the decisions; he or she wants the members of the organization to make decisions. To ensure consistency, the CEO communicates criteria to be used for decision-making, and these criteria are equivalent to the strategy.

This concept is best illustrated when one considers the approach of centralized command demonstrated by historic military leaders when conducting their campaigns. In their world, each individual soldier comprises a single fighting unit, as well as being a member of a ten-unit squad. Normally ten squads comprised a company, ten companies comprised a battalion, and ten battalions formed a brigade. By virtue of this simple architecture, the commander is able disseminate a command throughout a large body of soldiers with relative ease. By communicating his direction or focus to the ten individuals under his direct command, who then, in turn, communicated it to the ten individuals under their command, an order could be disseminated to several thousand soldiers in a very short period of time.

One difficulty that arose with this approach, however, is related to the telephone game often played by children. In this game, a phrase is whispered into the ear of the player next to the first. By the time the phrase has been whispered in the ear of five or ten players in the game, it becomes largely unrecognizable due to the limitations and challenges of communications. If we extend this limitation to the previous example, it becomes apparent that the specific details of the communication might not be as readily communicated as the spirit of the communication. This is in essence what strategy is intended to convey. Therefore, upper management must focus very specifically on the development of an appropriate vision and strategy that will be disseminated (shared) amongst the organization, the intent being that the spirit of the decisions is communicated. This will effectively establish for the associates an idea of "how" they should approach their decisions and what conclusions they should likely draw from them.

The lowest level of responsibility is that of the workers. The workers are typically located at the bottom level of the organization, and comprise the largest percentage of the population. They do not direct other workers in their efforts, and are generally responsible only for their own contributions. In many corporations, this is referred to as the individual contributor. As might be expected, the responsibilities of the individual contributor are to maximize their contribution to the organization, which is largely based on a task orientation. Naturally, this task orientation should be based on the strategy of the organization. In other words, the individual contributors support the organizational strategy by means of their decision-making throughout the business day. Therefore, the communication of the vision and strategy must be clear enough to allow appropriate interpretation by the members of the workers' group so that the leadership is assured decisions are consistent and coherent throughout all levels of the organization.

The remaining level of responsibility is that of middle management. These individuals are typically responsible for the development of projects and programs within the organization, and may also comprise the functional management of individual contributors within the organization. One is in a middle management role if he or she is making decisions that directly impact the discretionary spending of the organization in any manner that extends beyond his or her own compensation. Most project managers fall into this category, and virtually all of engineering operates at this level. In particular, the project leads play a middle management role within the organization. As has been previously clarified, upper management's role is one of developing vision and strategy, while the workers are responsible for task-oriented efforts. Clearly the ranks of middle managers must be responsible for the connection between strategy and tasks. This implementation of the strategy is known as tactics. Thus, the middle manager must define tactics in a manner that will support the organization's strategy while simultaneously outlining how tasks should be conducted, thereby accomplishing the necessary work of the organization.

To that end, we are interested in viewing the operation of the organization with regard to these areas of responsibility. This allows us to define a specific construct of responsibilities, known as the Goal – Strategy – Tactics – Tasks framework. This framework is defined in order to illustrate the challenges that most organizations have with the appropriate implementation of strategy. This can be seen in Figure 2.

To further convolute this issue, often the organization is split into vertical sections, or functional divisions (see **Figure 3**.) This will result in further separation of interests, often creating operational islands (see **Figure 4**.) These operational islands can be very problematic within the organization as they allow individuals within the groups an excuse to limit communication across the boundaries. While this typically is not done in a malicious manner, often the result is a costly waste of resources.

As an example, most employees in departments within an organization feel a certain affinity to their coworkers and a certain separation from those outside their department. This separation can be manifest in a variety of ways, and can include a reluctance to speak face-to-face, a limitation in communications via text or e-mail, or a perspective of an *Us versus Them* mentality within the organization. This very rarely increases the efficiency or effectiveness of products.

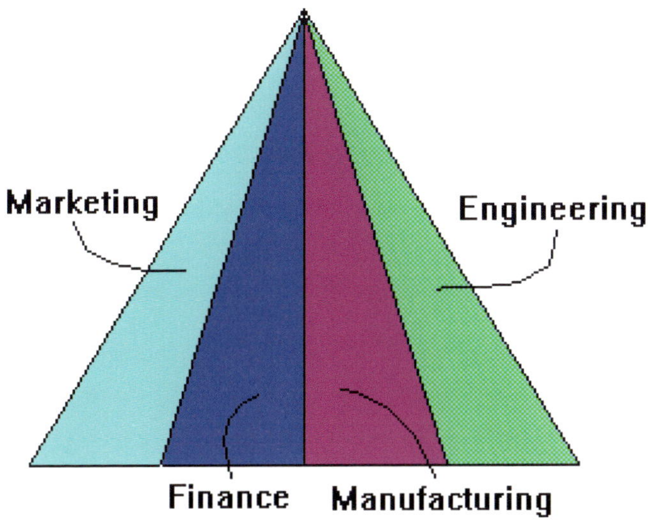

Figure 3: Functional Division

Management gaps + Functional gaps = Operational Islands => Dysfunction

Figure 4: Operational Islands

The value of the Goal – Strategy – Tactics – Tasks framework enables us to focus our efforts effectively within the organization. With proper execution, we can be sure that we have established a coherent and cohesive vision that enables the maximization of resources as they are assigned to tasks.

Imagine we want to apply this construction to a simple football game. When asked to define the goal of a football game, most individuals will respond that the idea is to win. While this is an important consideration, it is a less tangible definition that might be desired in the statement of goal. For example, when it is suggested that our real goal is to score more points than the opposition during the game, most people will agree that that is a more tangible and practical description of what needs to be accomplished by the football team.

If one considers the individual tasks within a football game, it is readily apparent that individual contributors are responsible to throw a specific block, tackle a specific player, run a pass route, throw a pass, run through the line with the ball, and many other specific tasks. These tasks are assigned individually to different members of the football team, often without direct knowledge of what all the other assignments may be. The tasks are assigned with the intention that a coordinated whole is created from these individual efforts. This coordinated whole is at the level of tactics in the football game, and it is equivalent to the offensive and defensive plays that are run during a football game.

The focus of the design of a play is to assign a wide variety of tasks to individuals on the field who are well suited to accomplish them, and then to orchestrate the whole in a manner that will allow the system to support the goal of winning the football game. With that said, it is clear that to this point we have not discussed an appropriate strategy for a football game. This is largely due to the fact that most people would approach the football game just as we have defined, without significant regard for the desired strategy of the football team.

It is of significant value, however, to consider the strategic focus for the football team. For example, if a team is highly successful at executing long passes, the team should choose plays that would enable them to throw a long pass. Likewise, if a team runs the ball well, the team will choose to develop plays that would enable them to run the ball on a frequent basis. These groups of skills or talents are often collectively referred to within our organizations as ‘core competencies’. Therefore, when considering the strategy, one should consider the core competencies within the organization.

So in consideration of our football game, one should approach the goal, to win the game, from the perspective of what the players are most skilled at doing. If we as a team truly understand our strengths, it would make sense to exploit them while minimizing potential weaknesses. So we must consider what strengths we bring to the table; the specific skills or competencies that we possess that will enable us to further our goal.

Once we have defined our desired focus for the game, we have essentially established the strategy that will guide our tactical decisions. For example, imagine that we are highly skilled at running the football. We would establish a strategy for our team that would enable us to utilize our core competencies and therefore more readily play our best game rather than the other team's game. This might include focusing on a short pass attack as well as a running attack, which would enable the application of core competencies and therefore allow us to take advantage of our finest points.

To sum up our approach, we have defined a goal of scoring more points than the opposition, and in order to accomplish this we have chosen to focus on our running skills (strategy) in the football match. When we need to design a play (tactic) for our team, we will take into account our strengths and competencies in order to design plays that enable us to use those strengths and competencies most readily during the execution of the game. Therefore, once the plays have been defined they will be disseminated to the individual contributors in a manner that communicates to them not only their specific assignments on the field but also the reasoning behind those assignments. It is always true that an individual contributor who understands the "why" of an assignment will more readily adapt that assignment to the best interests of the organization. If individuals are not given task justification, they have little or no idea how to adjust their tasks within a dynamic environment.

If we were to consider a retail example, one could consider an organization such as Wal-Mart. Walking through the same thought process, we define the same elements:

Goal: to be a profitable business
Strategy: to be the low-cost provider (Why the customer values you)
Tactics: to focus on cost reduction in everything we do

The most important of these defined elements is the goal. Often, goals are expressed in a manner that does not allow the individuals to easily interpret the goal for their specific assignments. In this particular case, we have chosen to define our goal as one of remaining profitable, which enables those who interpret this goal to choose a perspective based upon their contribution to the organization. If we had suggested the goal was to increase market share, this would be much more difficult to distill down to the assignments within the company.

It is clear that Wal-Mart approaches their efforts with a focus on cost. How do we know this to be the case? One need only consider why an individual would choose to go to Wal-Mart to purchase retail items. Most consumers would readily admit that they go to Wal-Mart for one reason alone, and that is to be sure that they are paying the lowest price possible for their goods. Therefore, Wal-Mart chooses to implement a strategy of cost. You will note that the strategy is defined in terms of the value brought to the customer. Another way to consider the strategy of any organization is to ask, “Why is the customer coming to me?” Defining the reasoning that makes the organization attractive to the consumer establishes the strategy that the organization should pursue and enhance.

In defining tactics for the organization or in defining those projects that will be implemented, we must consider this defined strategy of cost. In order to illustrate this point, we will imagine that we have been assigned the task of laying out the new floor plan for a Wal-Mart store in our locale. The only thing we know as a constraint in our design is that our primary strategy is to maintain the lowest possible cost to the customer. What would this floor plan look like and how would we define the characteristics of it?

To answer this question, we must consider what actions we could take that would potentially contribute to low cost. We should then select those actions that contribute most effectively to cost; those that have the highest return on investment. We would then implement those actions as our primary tactics. For example:

The floor plan will be large – in order to take advantage of economies of scale
The stock will be located on the floor – in order to minimize material handling costs
The stock will be delivered and placed on the shelves by the vendors – to reduce material handling costs
Stocking will occur during times when customers are not in the store – to minimize impact on sales opportunities
We must design multiple loading docks, and wide aisles for

the transport of goods within the floor plan by forklift or pallet jack – to minimize inefficiencies when multiple vendors are restocking items

As can be seen in this simple example, we can make our decisions utilizing the criteria of strategy in order to maintain a coherent application of resources within the organization. As many of us know from observation, this is what a Wal-Mart floor plan looks like, and now we understand why.

The Strategy of the Organization

To understand strategy, we can consider the definitions of the words.

Strategy –
The science and art of employing the political, economic, psychological and military forces of a nation or group of nations to afford the maximum support of adopted policies in peace or in war; the science and art of military command exercised to meet the enemy in combat under advantageous conditions; a careful plan or method, the art of devising and employing plans towards a goal.

Tactics –
Of or relating to arrangements of order; a method of applying forces in combat; a device for accomplishing an end, showing orientation or movement directed by a specific force or agent.

These definitions fit well with our previously discussed notion of strategy as it relates to maximizing the core competencies, specifically those competencies that your consumer or customer finds most attractive in your organization.

Often there is confusion relating to the definition of business strategy as well as product strategy within the organization. In the former case, we are defining the approach we choose to take toward the customer, and the reputation we seek from that customer based upon our actions. The business strategy can typically be described with a simple taxonomy as shown in **Figure 5**.

Figure 5: Business Strategy

In this example, four typical business strategies are defined and juxtaposed with regard one to another. They include Customer Service, Manufacturing Efficiency, Customer Intimacy and Product Leadership. They are arranged based upon the competitive values framework. This framework contrasts an organization's competitive interests, defining the inward/outward focus and the product/process focus present in the organizational culture. From this cultural taxonomy, common business strategies can be defined. Each of these business strategies will be discussed individually.

In their 1997 text, "The Discipline of Market Leaders," the authors describe three of these four focus areas, excluding service excellence in their approach.

Service excellence is a strategy specifically focused on satisfying customer regardless of the challenges faced by that customer from either the product or the organization. The main focus of the organization in this case is rapid response to customers, and the ready acceptance of responsibility for any shortcomings of the product is paramount. The strategy of service excellence is perfect for markets where customers value dependability of the equipment, and when that dependability is in question, support from the organization is provided to back its equipment. Leaders in this domain typically focus on having highly dynamic response teams associated with the equipment, and may limit their operations in order to maximize this capability.

Manufacturing Efficiency is a strategy that focuses on cost reduction. The typical organization applies resources to enhance its manufacturing capabilities with regard to automation and streamlining of work in order to produce the highest degree of manufacturing excellence. This strategy is applied by organizations that are focused on quality, as their product strategies require a high-volume product with highly standardized features and very little variation from those standards. This type of strategy of manufacturing excellence is ideal for markets where customers focus on cost only, which is typically the case for commodity markets and well-established technologies. This is manifest in highly standardized processes and procedures within an organization and a bureaucratic approach to making decisions.

Customer Intimacy focuses on providing whatever unique capability a customer may require for the product being offered. Often, this will result in highly customized or a highly personalized product for the customer. An alternate approach is demonstrated by organizations that consistently identify the future needs of their customers, prior to the customers understanding that need. This is often defined by organizations that offer bundled solutions to problems. As might be expected, pursuing this strategy requires that the organization have a deep understanding of customer needs as well as insights into future developments that apply to these needs. This also implies that these solutions are typically not the most innovative or the most cost-effective. A dynamic development environment and an ability to rapidly adjust to changes in the marketplace characterize an organization embracing this strategy.

Product Leadership is a common strategy employed by technology companies. This strategy is focused on possessing the superior product within the marketplace at any given time. Product leaders often can command a premium price for their product prior to commoditization of said product. Companies that utilize this approach must focus carefully on their ability to be innovative, problem solve and produce highly efficient teams. It is typically the diversity of these teams that yields the greatest results for these organizations. Therefore, companies that focus on product leadership also establish effective research management, product management and talent management programs in order to produce the visionary and dynamic workforce required to accomplish this challenging task.

While these business strategies are of significant use in focusing the efforts of organizational development, they can be challenging to apply to the individual contributor within the organization. For example, if your organization selects service excellence as a business strategy, what criteria will a Design Engineer utilize to select a bearing that provides the customer with the most effective service excellence?

Figure 6: Merging Business Strategies

Since it is unlikely that we will see behaviors in our organizations that require significant intellectual application or arduous tasks, such as deep thinking and interpretation of complex relationships, we must define our strategy in terms that require very little effort to interpret and apply on a daily basis. In short, our strategy must define the criteria for decision-making at the lowest levels of the organization. By limiting the breadth of interpretation available to the individual contributor, we ensure that all of the individual contributors are working coherently within the organization. This enhances our ability to compete by more efficiently utilizing resources in reducing or eliminating those squandered by misinterpretations of the strategy.

Strategy therefore provides us alignment for our team and compelling arguments for the wisdom of our direction. These compelling arguments are often very useful for the individual contributors, as they build motivation and engagement at the employee level. By having the engagement on our projects aligned with an appropriate strategy, we enhance the coherence of our efforts and assure ourselves that everyone is moving in the same direction.

Strategy essentially becomes our:

Guiding principles
Ingrained behaviors
The way we do things
Criterion for all decisions

A simple taxonomy of competitive strategies is utilized to simplify the task of strategic development. Wheelwright informs us that there are only five principal strategies typically implemented by organizations. These are known as competitive priorities or product strategies, and describe what the customer wants in your product.

Competitive Priorities

The Competitive posture in key areas of the business should focus on one primary strategy and no more than one secondary strategy. The organization needs to be reasonably competitive in all of these areas, with a particular strength or capability in the one primary focus area. This strength defines the core competencies. The five product strategies include:

1) Price	(Cost)
2) Dependability	(High Value)
3) Quality	(High Consistency)
4) Flexibility	(Choice)
5) Innovation	(First of a kind)

A company cannot usually succeed at all strategies simultaneously and must choose one or two and attempt to excel in those areas, or it risks coming in at second place in the competitive arena.

A competitive strategy of price is normally associated with a focus on cost for the customer. This strategy is most commonly associated with organizations focused on high-volume production and who are able to establish economies of scale. All decisions focus on the bottom line for the consumer, which may also imply limited margins and profitability for organizations choosing to adopt this strategy.

A competitive strategy focused on dependability or reliability is normally associated with high-value products. Therefore, the companies that utilize the strategy typically provide higher cost products within the marketplace than others' readily available products. These products typically perform at a higher level, or for a longer time, or better (in some way) for the customer. It is critical for these organizations to define that they are providing excellent value for the money for the customer. As an example, if you produce a pencil that lasts ten times as long as a normal pencil, presumably you can charge a premium price as it produces more value for the customer than a normally priced pencil.

Organizations that select quality as a competitive strategy are focused principally on high consistency in their product offerings. This is normally associated with mass production, and aligns well with a Manufacturing Excellence business strategy. Companies that utilize this strategy focus on highly repeatable products with very few changes, modifications or defects in the manufacturing process. By driving out both waste and non-value added activities, these companies can provide significant value to their customer.

Flexibility is a competitive strategy chosen by organizations that want to provide their customers with a great deal of choice. This may be illustrated by a wide variety of product models or a large option list associated with their products. This strategic choice is often of value to customers, as they are seeking highly personalized or highly focused solutions for their needs. Companies that focus on flexibility must also demonstrate an ability to rapidly adjust their product offering to the needs of a dynamic customer base. This solution often results in a higher cost to the customer and higher operating costs for the organization.

Innovation is an effective strategy selected by organizations that focus on creativity and rapid product development. Their products are often characterized by being the first of their kind in the market, new and exciting technologies, and products that cannot be obtained from any other organization. As might be expected, companies that adopt this strategy typically have high development budgets and more long-term technology development plans than organizations pursuing other strategies. They also depend on their customers' willingness to pay significantly more for an innovative product.

Implications for Engineering Leaders

It is imperative for engineering leaders to spend time considering the various product strategies that may be implemented by their organizations. Although the appropriate strategy may not always be clear, it is incumbent upon engineering leaders to define appropriate product strategies and disseminate that understanding. A failure to do so will result in a potentially unfocused development approach, which will most likely be accompanied by failed products and projects.

One technique that has proven successful is to utilize a stakeholder analysis with regard to potential products and projects within the organization. A sample stakeholder analysis worksheet is supplied for your consideration.

Stakeholder Analysis Worksheet

Stakeholder	Overall Interest(s) in Project (Spec/ Sched/ $)	What we need from Stakeholder and When	What Stakeholders want from Us and When	Key Focus for Actions

Integrating the Project Management System

Any Project Integration project can be either a nightmare or a blessing, depending on how well it is planned and executed. This becomes more the case if the project is extremely complex, or involves disparate applications and systems and has to be executed in conjunction with third parties. Under conditions such as these, it becomes extremely challenging to deliver the system within the desired timelines and at the required performance level. The integration of a complex system such as that described in this text is just such an application.

In order to accomplish the integration of disparate concepts, one might consider approaching this task as a System Development project. Thus, many of the concepts presented in the text can be directly applied in order to accomplish the integration of the system and processes within the engineering organization.

We would propose the following methodology, describing a staged process to accomplish the integration:

1) Define a leadership and review team with members drawn from the organization and any significant external parties necessary for the implementation
2) Establish the project parameters, in terms of defining the purpose and corresponding and deliverables as well as any significant time constraints
3) Plan and implement a detailed project schedule, including interdependencies and resources required for completion
4) Define a team structure and engage the team within the

project environment, and define what successful integration looks like in a common format

5) Establish the project priorities, in terms of selecting those methods and processes that the project will implements, as well as team communication needs and status for the project

6) Establish a business rhythm to ensure the project remains focused and on track for success

During the execution of the systems integration process, scope management will become a significant factor. Scope typically refers to all the work involved in creating the products of the project and the processes used to create them. A deliverable is a product produced as part of a project, such as hardware or software, planning documents, or meeting minutes. Scope management includes the processes involved in defining and controlling what is, or is not included in a project effort.

Scope management is the process by which we need to find a method of establishing the scope, defining the scope, verifying the scope, and controlling changes to the scope. Typically, this might be accomplished with a project charter document, which will define the scope statement as well as deliverables defined for the project. Additionally, the work breakdown structure will contain many elements related to scope content.

As time progresses, the scope of the project should become clearer and more specific. It is likely to become more refined based upon the increased knowledge and insights gained in the development. This is shown in Figure 7.

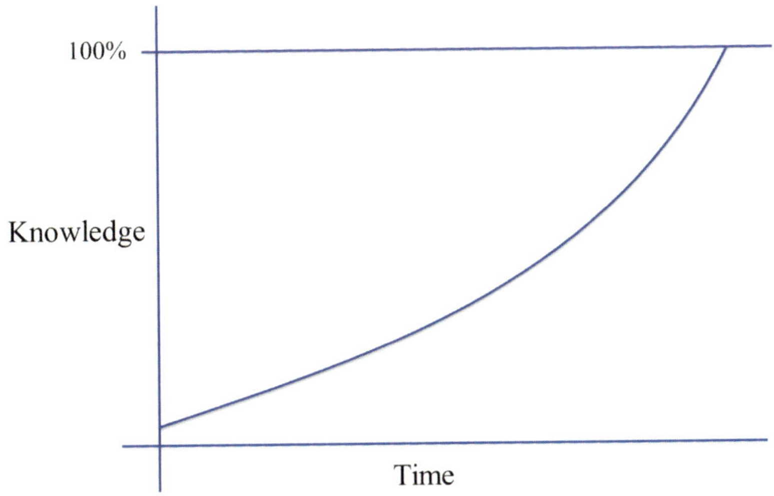

Figure 7: The Knowledge Curve

So why does this scope change significantly during the life of the project? First, it is important to recognize that the majority of planning for the project occurs very early in the life cycle of the project itself. Second, the end product of the project is typically the generation of knowledge sufficient to allow the release of the project for production or process utilization.

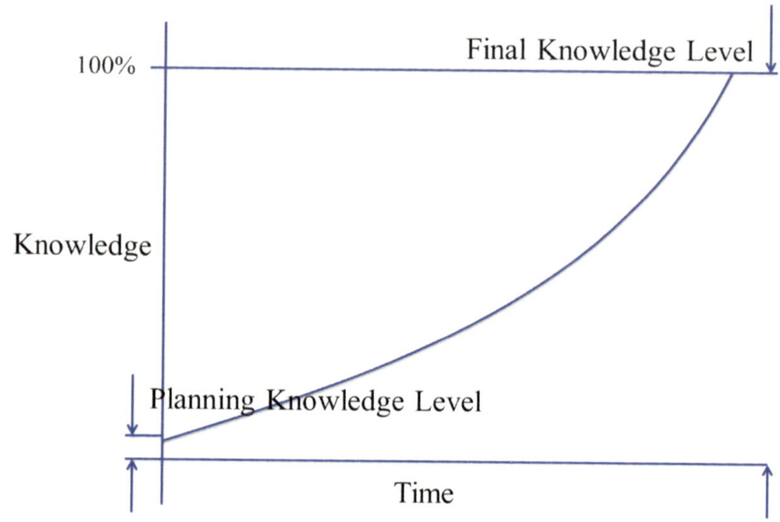

Figure 8: Why Project Scope Continually Changes

If one considers the level of knowledge developed at the outset of the project and compares it to the level of knowledge obtained by the completion of the project, it is apparent that many of the decisions and perspectives at the outset of the project may look less informed or totally incorrect by its completion. If this is true, then in any system's development the engineering leadership should fully expect significant scope changes. Therefore, scope control, which is often portrayed as the suppression of changes within the project, is actually the ability to appropriately manage scope change.

This implies that the engineering leader must be willing to accept the fact that projects and systems are exceedingly dynamic and will likely change on an ongoing basis during their development processes. This is also true for systems integration efforts. As we become more familiar with the challenges and opportunities presented by the integration of the system of this complexity, we must be willing to constantly modify and adjust our plan in order to ensure that the deliverables are topical and appropriate for the work investment.

Of greater importance to the engineering leader is effective change management. Change management is the process of developing a planned approach to implementing change in any organization. The objectives of change management are:

– Maximize collective efforts of stakeholders
– Minimize the risk of failure of implementation
– Ready the receiving organization for the results of our project

All projects incorporate change into your organization. Those that do not incorporate change are not necessary. By virtue of this, we must become active change agents within our organizations, as well as effective managers of the change as projects are implemented.

Although all people are unique, we can have similar responses to change. In general, we view changes that are forced upon us as unnecessary and irrelevant. Many perceive the inconveniences caused by change as malicious attempts to sabotage our work or focus. In reality, the change implemented is rarely intended with ill will, and instead is simply not managed properly.

Organizational Change Management is the discipline of helping people adopt and accept change within the organization. It helps transition individuals, teams, and organizations from a current state to a desired future state. Why is change management important? Ultimately, it enables timely project implementation in an engaged and ready workforce, and it avoids negative interruption to the business.

As an example, at GE Capital, one hundred percent of all projects evaluated as “successful” had a good technical solution. Interestingly, over ninety-eight percent of all the projects evaluated as “unsuccessful” also had a good technical solution.

Ultimately, change management involves the development of both stakeholder management, as well as change adoption processes within the organization. These two factors are defined as the quality and acceptance of the solution.

Stakeholder Management is the process by which a project’s key stakeholders are identified and an approach to win or maintain their support is developed. Good Stakeholder Management leads to quality project execution and deployment.

Change adoption is intended to enhance the organization’s readiness to accept change. The majority of this effort is to recognize that change in an organization requires time. In general, people will pass through varying reactions to the change as they process an implementation. These stages include:

Satisfaction
Denial
Resistance
Exploration
Hope
Commitment

Therefore, to manage the change adoption appropriately, we must follow a simple process. This includes:

Define why this change is needed
- Explain why the approach selected is optimal

Craft a business case
- Demonstrate or prove the business benefit

Create a compelling story for the change
- Make this story simple to understand and share

Align various stakeholders

- Focus on Consortium building
- Provide a constant flow of communications to get the organization accustomed to the change prior to the implementation

The engineering leader must understand that all engineering endeavors introduce significant change into the organization. The individuals to whom the change is introduced dramatically impact the success of the endeavor. As an example, in developing a significant process improvement for the manufacturing floor, you introduce significant change to the operators who are interested in producing consistency and quality in their daily efforts. In reality, the individuals who have been negatively impacted by the introduction of a change are the same individuals who will either make or break the project. Therefore, it is imperative for the project leader to focus on the Change Management process well in advance of the implementation of the change.

Phase 2: Project Planning

Project Planning is a crucial component of the project life cycle. There are many different published methods regarding planning, and several will be discussed here. Fundamentally, it comes down to a very simple understanding of what needs to be accomplished in the Project Planning process. There are three fundamental questions that must be answered:

What are you going to do?
How long is it going to take?
How much will it cost?

If these three questions are answered effectively, any individual can make an appropriate business decision based upon knowledge of the organization and this scoping of the project. Therefore, we will spend our time focusing on how to address these specific needs.

In order to answer the question, "What are you going to do?" we must scope the project appropriately, provide requirements that must be fulfilled for each of the appropriate stakeholders and ultimately derive a specification describing the fit and function of the system being developed.

The specification will describe every aspect of the system under development, but we also want to identify the higher-level deliverables that must be provided by the system in order to provide a complete description of what will be accomplished on the project.

There are several approaches that may be used to develop descriptions of what will be accomplished, but prior to that a significant effort that must be undertaken to define requirements necessary to achieve all of the stakeholders' needs under the project development effort.

Defining Users

When beginning a project, the project manager must understand who is interested in the outcome of the project. Those individuals and groups are often described as users or stakeholders for the project. The project manager will need to identify all of these users in order to begin compiling and understanding their user needs.

In general, there are similar stakeholders on virtually every project. These will include:

Supervisors
Executives
Sponsors
Project team members
Customers
Suppliers
Competitors
Other department members
Subcontractors
Consultants

If we take a simple example of the development of a new coffee maker for our company to offer as a product, we might consider the following users:

The project team members

The project manager
The sponsors of the project
The potential customers we are identifying
Coffee suppliers
Retailers
Shippers
Recyclers
Packagers
Marketing
Sales
Finance
Manufacturing
Purchasing
Safety
And the list goes on...

Project managers must create an exhaustive list, including all the stakeholders that they would like to take into account with regard to their development effort. Obviously, many of these stakeholders will hold contradictory or impossible needs that they feel the project should accomplish. It is important for the project manager to at first include all of these stakeholders, although by the completion of the project they may have elected to remove some of these users from the list. This often happens when it is unrealistic or impractical to address the needs of specific user groups or individuals.

Determining User Needs

Once the various users have been identified, the development team's project manager will spend considerable time developing an understanding of the needs and desires expressed by these various users.

These usernames will typically be expressed in language familiar to the user, although this will not translate directly amongst all of the users or even amidst the project development team. Therefore, significant effort is required to translate these user needs into useful information for the planned development effort.

Ultimately, the development team will translate user needs into the requirements on which the project must deliver in order to be deemed successful. For that reason, the development team must make significant efforts to fundamentally understand what capabilities and functions must be delivered by the product in order to satisfy the majority of the users and their expressed needs. In addition to this, it may be of equal importance to identify and address unexpressed needs that can be observed by the development team.

This results in a multiple-stage process, which requires the development team to elicit needs from the users, translate them into practical requirements, and derive additional requirements for the project in order to establish what criteria must be met by the final delivered product.

Requirements Definition

A requirement is something that the product must do or a quality that the product must have. A requirement exists either because the type of product demands certain functions or qualities, or the client wants that requirement to be part of the delivered product.

Why focus on requirements?

- To ensure project success, 8% - 14% of project costs should be invested in requirements development
- Requirements define the project scope
- Make / Buy decisions can be made based upon the requirements
- Requirements can define how we implement the project
- Requirements reveal where further research is necessary
- Requirements help define what we do not know
- This allows the requirements to drive the solution - not the reverse

Reasons for Good Requirements

- Save time and money in project development
- Increase efficiency of communications
- Set appropriate expectations and parameters for the project
- Prevent duplication of effort by project team members
- Focus project resources on key milestones and maintain coherence
- Spend less time on revising product or service requirements during or after project implementation
- Fewer contract changes with external vendors and associated costs
- Better product or service quality
- Performance accountability – to be able to do what we said we would do

Many different methods have been defined to enable us to begin to develop requirements. In general, there is no ideal method that will ensure that all the requirements or even the most critical requirements have been identified. Utilizing multiple techniques such as those described in the following paragraphs will be very useful in establishing a large percentage of the requirements that will be of interest to the project team and customer.

Requirement Definition Techniques

Interviews/Surveys/Focus Groups
This method is typically used when little is known about the expectations and needs of the customers and stakeholders. The basic premise of this methodology is to talk with people utilizing a structured interview technique in order to define:

What do they need from the developed project?
What do they want from the developed project?
What are they concerned about with regard to the effort?
What other elements of interest come to mind when they are considering the project?
What capabilities are implied that must be present in a successful system?

This methodology can be quite time-consuming, but allows its user to discover interesting perspectives and innovations that might not readily come to mind without the input of those most familiar with the application, the stakeholders.

Mind Mapping

Mind Mapping is a methodology that begins with our project focus located in the center of the mind map. From that center, we connect all related topics that might be of interest to us with regard to the project. This allows us to create a series of spokes around our central concept that will bring to mind many considerations we might not have perceived prior to this exercise. In addition to this, we can continue to connect spokes outward from the related ideas to more fully develop an understanding of what challenges need to be addressed in the project. This may be seen in **Figure 9**.

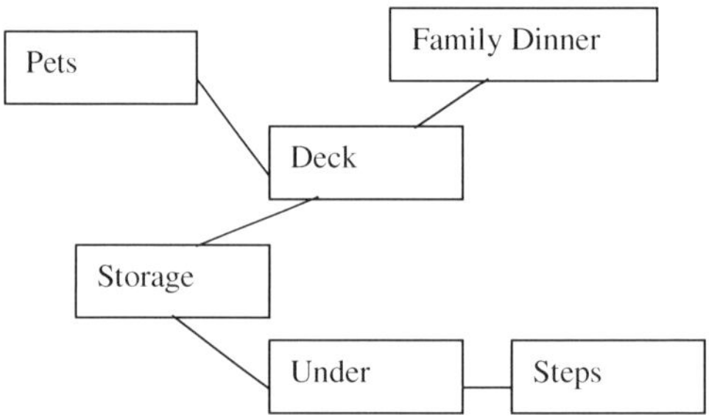

Figure 9: Mind Map of a Deck Project

Workflow Analysis

In workflow analysis, establishing the workflow throughout the life cycle of the system being developed generates its requirements. This will essentially result in a Start – to – Finish Block Diagram that will be used to operate the system. From this diagram, a developer may carefully examine the relationships and establish whether any of the steps may be eliminated or if any challenges are present in the current workflow.

Additionally, this workflow can be connected to other systems within the organization, allowing the developer to explore the relationships amongst these interrelated systems. From the workflow analysis, each box and each connector will typically result in one or more requirements being identified.

A Cake Example produces a workflow:

Decide on Cake
Find Recipes
Check Inventories
Go to Store
Buy Groceries
Prepare Kitchen Components
Mix Cake
Grease Pan
Bake Cake
Cool Cake
Make Frosting
Frost Cake
Serve

From this example, it becomes clear that the project plan must include recipes and buying groceries in order to complete the overall project. This type of workflow analysis may help to create boundaries around the project enabling its stakeholders to have a common understanding of what will be accomplished and what will not be accomplished within the project development.

Systems Analysis

Systems analysis is commonly used in complex systems development. The preferred method typically utilizes an object-oriented model of the system functionality in order to fully understand the interrelationships of the various elements of the system as well as the capabilities that must be delivered in the final product. This is similar to dataflow modeling in software development or event response modeling as is commonly used in electrical engineering.

As further detail is added to the system model, each of the boxes may be expanded into multiple functions that generate additional understanding and requirements. In the Systems Engineering field, this type of systems analysis is what generates the lion's share of requirements for the system developer. Further details on this technique can be found in the Systems Engineering section of this text. An example of a top-level systems analysis is shown in **Figure 10**.

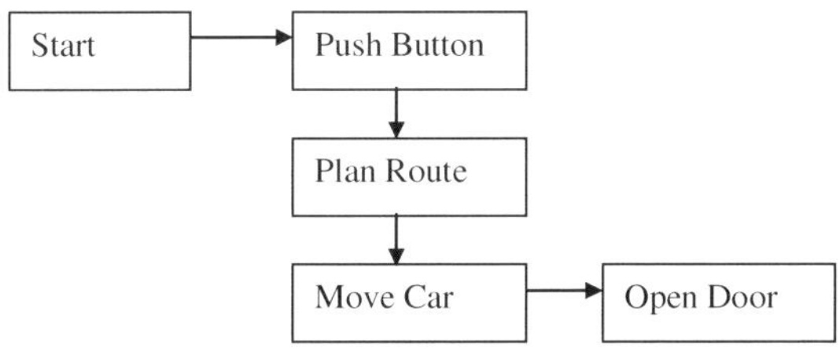

Figure 10: Systems Analysis of an Elevator

Prototyping

Prototyping is undertaken to address one of the fundamental shortcomings of customers in the project development process. The problem is in specifically identifying what thy need or want from the project. Often, this is due to an inability to see what is currently not existent, and it limits the customers' perceptions only to what they already have. The use of prototyping can generate new products and new approaches that will enable customers to better understand what you intend to accomplish and may provide additional requirements based upon their observations regarding the prototype.

The prototype can be a display-oriented or functionally-oriented device, and can be rudimentary or elaborate depending on the technology availability and the criticality of the prototype results. In some cases extremely high fidelity prototypes can be created in CAD systems in order to provide the customer a realistic view of what the finished product will look like.

Additionally, prototypes may be used where modeling or simulation could become too demanding or expensive in the development of a usable end product.

Scenarios / Use Cases

Scenarios are activity-based descriptions or models of the system's intended use. Use cases are similar to scenarios except that they also include the actors associated with the activities. Both of these approaches may be very useful in understanding the interrelationships between the systems as well as between the actors in the system in a typical scenario. Additional insights and requirements may be generated for the system by walking through all of the activities and relationships.

Market Research

It is often helpful for the project leader to conduct significant research into the market or the project's technological area of interest. This research might take the form of:

Internet searches
Publications review
Reviews and discussions with business contacts

Many project leaders are surprised to discover that in many cases other parties, sometimes in unrelated industries, have already accomplished the work they are contemplating. Some research is typically required to better understand the challenges faced by the project team prior to project launch.

Templates/Project Warehouses

In many cases it is well worthwhile to examine past efforts undertaken by the organization or records available in the public domain from other organizations that have accomplished similar projects. These documents can describe standard requirements that are utilized in that industry, and save considerable time in requirements development. These documents might include:

RFP – request for proposal*
Project introduction/kickoff documents
Business objectives
Background information supplied with the project
Scope descriptions from project sponsors
Business terms and conditions from contracting agents or vendors
Requirements documents included in business agreements
Proposal instructions included in an RFP
Evaluation criteria included in an RFP

*Note that this document is usually not available at the project level; this is more of a program level activity.

Performance Based Requirements

Performance-based requirements are typically the most useful for the project development team. These are requirements that are measurable and achievable where the performance measure is included in the description of the requirement. This requires that care be taken in the development of requirements to ensure the resulting requirement description is appropriately constructed and has enough detail to provide a performance-based profile.

Several rules are utilized to generate effective requirements. These include:
1) The requirement statement is a complete sentence.
2) The requirement statement uses the word 'shall' in its construction.
3) The requirement is limited to a single thought.

Example performance-based requirements will focus on issues such as cost, quality, speed, efficiency, weight, and size, to name a few.

Example performance-based requirements might include:
General: Must print fast.
Performance-Based: Must print three inches per second.
General: The software system shall retain history.
Performance-Based: The system shall retain a minimum of sixty weeks of history and a minimum of four weeks of schedules.
General: Training shall be provided.
Performance-Based: Three training sessions shall be provided to thirty employees onsite at XYZ Corporation.

Types of Requirements include:

Functional - What does the product or service need to do?
Action requirements
Workflow steps

Examples:
Sum total hours worked
Calculate payroll
Print paychecks

Technical - Conformance to standards
Conformance to IEEE 802.11(b) Standard for radio frequency
System type
900 MHz vs. 2.4 GHz direct sequence spread spectrum (DSSS) system.

Usability - Describes how the functional requirement should be perceived:
User-Friendly
Graphic user interface
Wizard-driven screens
Reliability: Mean-time between failures
Speed: Number of seconds/minutes/hours to complete
Accuracy: Percent accuracy or allowable errors
Number of steps to complete
Reliability: Mean-time between failures
Speed: Number of seconds/minutes/hours to complete
Accuracy: Percent accuracy or allowable errors
Number of steps to complete

Operational/Environmental - Describe the system environment:

Components must tolerate external temperatures from –20 degrees to 110 degrees F.
Equipment must withstand 95% humidity without deterioration in performance as measured by MTBF

Security
Confidentiality
Encryption, authorization, authentication; privacy policies
Accessibility
Legal considerations
Physical Characteristics

Equipment size limitations
Portability
Weight limitations
Materials – durability
Company brand requirements

Support
Maintainability
Instruction manuals if to be maintained internally
Knowledge articles for internal support centers
Spare parts
Developer kits
Service Level Requirements
Repair turn-around (e.g. Must be repaired within 24 hours of service call)
Service representative availability (e.g. 24x7)

Training
Train the trainer
Train specified number of business or technical users
Describe level of participant skills
Number of hours/days
Specify content to be covered at a high level

It is generally considered good practice to ensure that some requirements appear in each of these specific areas after requirements definition is complete. Once all appropriate requirements have been defined, a completed statement of work and detailed specification can be completed. An example statement of work document has been included for reference at the end of this section.

Developing the Specification

The Need for Integrity

To fully understand the value of coherence in our project efforts, we must spend a bit of time considering the topic of integrity. This is important because it enables us to place in context the importance of focus incoherence for our project team and their subsequent efforts on our project development. For this reason we will consider a number of interesting questions:

What is integrity?
What does it mean?
Why is it important to us as project managers?

Initial consideration of these questions often leads to discussions regarding honesty and morals within the workplace. However, when one examines the complete definition of the term integrity, one sees that there are other interesting facets of the definition that might be perfectly appropriate to consider in our project planning. A simplified version of this definition is as follows:

Integrity

1) The state of being whole or entire.
2) A sound, complete, or unimpaired condition.
3) Uncompromising adherence to one's principles; soundness of character, honesty.

When we consider our project with regard to this definition, we realize that there is significant importance in maintaining a sense of wholeness or completeness. In addition to this, there is significant importance in demonstrating an uncompromising adherence to principles. When we consider how this applies directly to our project efforts, we must justify the appropriate principles to be used in the decision-making for our project, and we must communicate these principles to our project team and stakeholders such that they will utilize them in their day-to-day activities.

This requires us to reflect on what principles or values will be of significance to us in our project development, and which of those principles should be used as guiding factors in the decision-making process. This will require significant thoughts and reflection, but the expected end result will be a provision of commonality in decision-making on our project, which will be the creation of coherence in our aggregated work.

Aristotle described the term rhetoric as the art of persuasion. In fact, he suggested that there are typically three methods that are utilized in persuasion, and these include:

Logos

A logical perspective or argument

Pathos

An emotional perspective or argument

Ethos

An ethical perspective or argument

Therefore, we can construct a compelling argument for our team regarding the importance of our project by utilizing these different approaches to persuasion. Consequently, there will be value in taking a few minutes and constructing a logical view of the importance of our project and emotional view of imports of our project, as well as an ethical view of the importance of the project. Consideration from these multiple perspectives will lead to a more detailed understanding and a more thorough consideration of the value of the project and the selected technical approach.

Why would this be important? As we have previously considered, human beings have a natural tendency to minimize their mental involvement in efforts that require significant reflection or intellectual investment. By virtue of this, many of the individuals who will perform on a given project are not going to spend a great deal of time considering why this project is crucial to the future of the department or organization. This should come as no surprise to a project manager.

However, the productivity of the individuals contributing to your project is typically governed by the considerations that the contributors themselves bring to the table. Plus, if the employees are frustrated with the organization they are likely to produce at a lower level of productivity than those who are more engaged. Any effective project manager understands that increased engagement in the project will result in increased productivity and a higher probability of project success. Therefore, it is incumbent on the project leader to ensure that the project team is fully engaged in the project. This will typically be accomplished by providing the project team with a compelling reason to apply their efforts effectively to the project development. This can be accomplished by following the typical model utilized by organizations to define their value statement, mission statement and business statements.

To begin, we will focus our efforts on the generation of a values statement. This statement will be useful for the project team in understanding what criteria should be used in their day-to-day decision-making. Therefore, we will consider a list of the values we expect our project or organization to demonstrate. These may include:

Honesty
Safety
Low-cost
High performance
Integrity
Strong customer focus
Service with a smile

While it is interesting to make a list that makes us feel good, we find a value statement must include the values that we choose to reflect in our business decisions. These are often not considered, and so we must reduce this list to the essential values of importance to us. Therefore, we will consider a slightly different approach. In this case we will attempt to address our values in two separate ways:

Define the value
Define why it is compelling for our business

As an example of this, we can consider me original DuPont Corporation of Wilmington, Delaware. The history of this organization is interesting, as it was originally a family-owned business that specialized in the production of explosives. When one spends a bit of time considering these two factors, it becomes obvious that safety would become a value for that organization. So we not only consider what the value is, but also why that value is critical to our object or organization.

From this list, we can prioritize our values and define a detailed value statement. For example, we may decide that we are managing a restaurant chain in United States. We begin to think about our values, and realize that our customers recognize the consistency of our restaurants and the safety they feel when they eat in our restaurants. Therefore, we define our value statement as:

We will provide meals to our patrons that are consistent in taste, appearance, and delivery time.

The mission of your organization is closely related to the emotional connection that your customer feels to the organization or its product. This is therefore synonymous with the strategic focus of your effort. This can also be thought of as considering what your customer wants from the project, and focusing on the characteristics with regard to the work, which will, in turn, create the most compelling argument for your project team. Therefore, one should consider the basic available strategies and select the one that best describes why your customer is interested in working with your organization to complete the project.

In the restaurant example we have previously described, our mission statement would read: we will provide meals consistent with our meal production process within a period of five minutes from order to delivery to the customer.

Finally, we would define logical arguments that become our business statement. This is typically the easiest of the three statements to generate, as it seems to make the most logical sense. However, we should leave this argument until last, as we find a more compelling statement usually resides in the mission or values for our organization or product.

In our restaurant example, this logical argument would generate the following business statement: by providing consistent meals in a rapid delivery format we will maintain and increase our customer base.

The utilization of these three statements to create a compelling argument is the basis that will create a coherent focus for your project team members. This coherence provides the integrity of our project approach, and allows us to more efficiently and effectively complete our work, as waste is minimized on those things of secondary importance to the project. Additionally, our workforce will be engaged by the compelling argument and will therefore produce at a higher level. If one considers past project experiences, it is clear that the projects that provided a compelling reason for existence typically received a greater productivity from the project team members. This is of course the fundamental goal of our efforts at this stage of the project life cycle.

Feasibility Studies

At this stage, feasibility must be considered with regard to the developments under consideration. The objectives of a feasibility study includes finding out if a system that moment project can be done effectively, what is the final product that will benefit its users as expected, the possible alternatives that exist and whether anyone of these alternatives would be preferred to the proposed project.

There are typically four types of feasibility to be considered at the outset of the project:

Technical –

Do we know how to do this?
Is the project possible with current technology?
How much technical risk is there?
Can the technology be purchased?
Can the technology be developed effectively?
Will the final system be compatible with our existing systems?

Schedule –

Can we meet the window of opportunity?
Is it practical to attempt the proposed schedule?
Are there resources available to meet the schedule?
If we missed the window of opportunity, what is the impact?

Budget –

Will we be able to bear the cost of the development and the finished system?
What economic benefit will result from the system?
Does the system have an attractive cost-benefit analysis?
What are the total life cycle costs?
Do the benefits outweigh the costs?
Is the project possible given our existing resource limitations?

Operational –

If we build it, will they come?
What will be the ultimate acceptability of the solution?
Will the proposed system potentially meet the users needs?

Will the organization support it?
Will society accept it?
Are there legal or governmental regulations that will impact this project?

The feasibility study is typically a managerial activity, and does not require a great deal of input from the project team members. It is, however, a crucial activity that requires the project leader to be completely honest and open about the probability of success with regard to the project, as well as if the project will fulfill the stakeholders' needs effectively. The project manager should consider the feasibility study as the first off-ramp for the project. In the event it appears infeasible, it makes perfect sense at this point to reshape the project or consider canceling the effort.

Statement of Work Worksheet

Project Title:	
Project Leader:	Date Written:
Project Purpose:	
Background:	
Project Description:	

Deliverables:
Measures of Success:

Alternate Statement of Work Worksheet

Project
Name:____

Project Mission
(strategy):______________________________________

____

Objective Statement:

Specification:

Project Specifications

A detailed specification conveys to all interested parties exactly what will be accomplished on the project, and the characteristics of its deliverables. This satisfies all the needs and requirements through the development project and will support the development of the top-down deliverables for the project.

The specification is intended to convey what will be done, and not describe how it will be done. In general, there is a very fine line between these two points, but it is important to draw a distinction between functionality and implementation.

Specifications should be as concise as possible, but should also effectively communicate the look and feel of the project deliverables, the functions or services provided by the deliverables, and the key performance parameters (KPP) for the project. The specifications should also convey the success criteria for the project.

To develop a specification one can utilize several techniques.

1) Consider the high-level project deliverables:

    Define each tangible and measurable component
    Define the components' notable characteristics
    Establish several requirements in each of the domain areas

2) Establish the functional hierarchy

    What is the top-level deliverable?
    From what subcomponents may this be made?
    What elements comprise the subcomponents?

Imagine we are defining a specification for a vacation we are planning in the coming year.

What are the deliverables?

An enjoyable vacation

Subcomponents of this include:

Travel to the site
Housing at the site
Meals
Entertainment/chores
Travel back home

Our specification should reflect each of these areas.

Our next task is to add additional details to each of these subcomponents as follows:

Travel to the site

Plan from home to the airport
Plan for a flight to the destination locale
Plan for transport from the airport to the destination
Established luggage plan for all of these
Consider meals as required during the transport

Housing

Plan for seven nights
Plan to remain at the same hotel
Make sure the lodging meets our standards

Once this is completed, we can continue to add additional detail in order to complete our specification.

Project will include travel to the destination site, including:

- Reserved cab ride to airport for a party of four
- Reserved flight from MSP to ABC for a party of four
- Reserve midsize rental car at ABC for eight days
- Luggage limited to one carry-on per member of party
- Luggage checked from MSP to ABC
- Meals purchased at MSP and ABC for party of four

Project will include housing at the destination, including:

- Two rooms of double occupancy at four-star hotel
- Seven-night stay
- Etcetera

Simply increasing the level of detail included in the description can develop a thorough specification. In general, detail should be added until the project plan becomes specific enough to be understood by the majority of its readers. While it is true that some of the individuals associated with the project may purposely misconstrue elements of the specification, it is unrealistic to expect an all-inclusive specification that will provide an airtight description of the project to be completed. The final project specification should take the form of a written document describing the requirements and specifications for the system to be developed. This may look like the following:

Requirements for a household toaster (partial list)

The toaster shall operate on 120 VAC 60 Hz power
The toaster shall allow for user adjustment
The toaster shall have a power cord utilizing an LG plug
The toaster shall be well tested
The toaster shall provide indicators for the user
The toaster shall have maintenance access panels
The toaster shall be portable

The toaster shall contain a safety override ground fault circuit interrupter

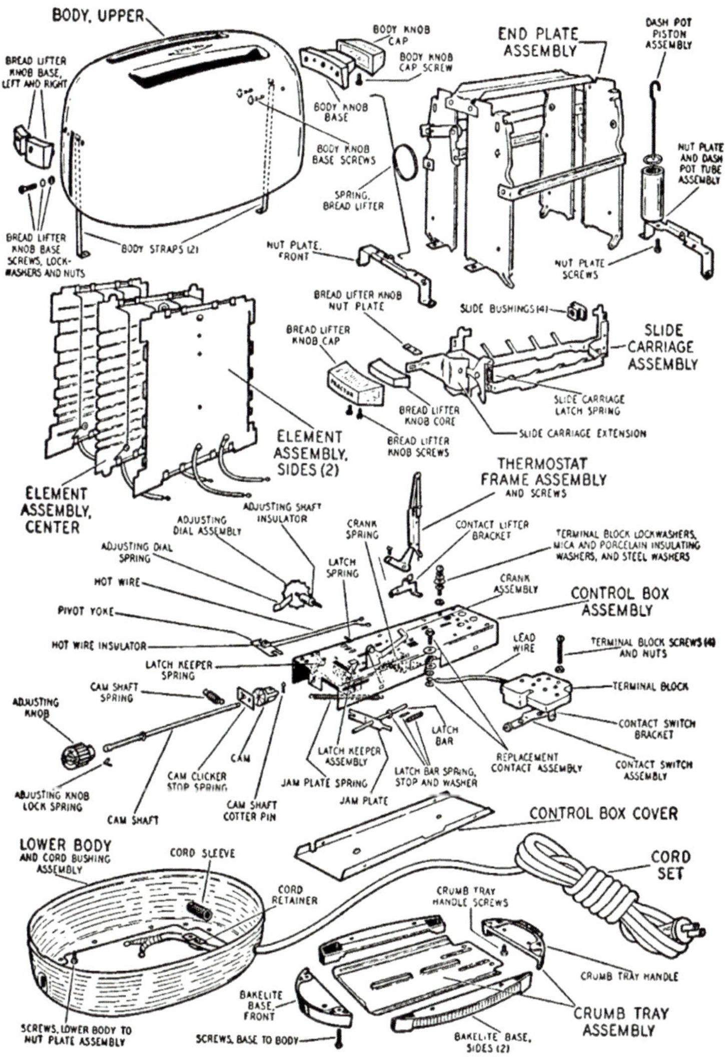

Figure 11: Graphical Toaster Specification

At this point we have defined one of the three factors of interest to the project manager; namely, "What we will do." Next we move on to the development of the project schedule.

Developing the Project Schedule

In order to develop a project schedule, project leaders must shift their thinking from the tangible deliverables of the project to a hierarchy of tasks that must be completed in order to accomplish the needs of the project. Different methods are suggested for the development of this hierarchy of tasks. The author prefers the approach of Functional Analysis to define not only the capabilities of the system, but also for the identification of the tasks that must be accomplished in order to produce those capabilities.

To conduct a Functional Analysis, the project planner must consider the operation of the system under development, and identify all needed capabilities by means of a verb/noun pairing. This is intended to provide a series of capabilities that must be assembled into a holistic view of how the system shall operate. This can be accomplished by a free association exercise in which all of the capabilities are identified, allowing time later in the development to fill in any missing functions or capabilities.

If we consider a simple example, the development of a countertop toaster for our home, one can imagine that we have identified a variety of capabilities that are communicated in this format:

Create set point
Clean cycle
Convert power
Update software
Override heat control
Shutdown power
Apply energy
Sense width
Toast product
Since complete
Shut down heat
Unload product
Loan product
Signal status
Input command
Request update
Reset manually
Clean device
Check load condition

These functions must then be assembled into a practical whole, which will provide an Operational View of the toaster system. This is shown in **Figure 12**.

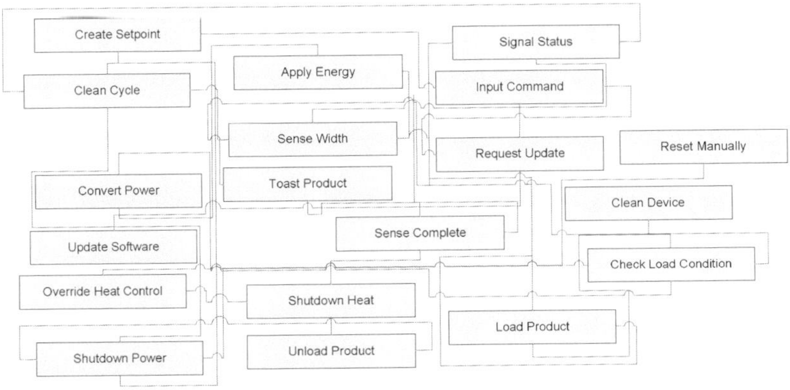

Figure 12: Functional Analysis of Toaster

Once all of the functions have been related one to another, the planner can untangle this diagram, producing what is generally constructed as a top left to bottom right flow of operations for the system. At this point it might become clear that there are several shortcomings that must be addressed, and additional functions may be added or superfluous functions may be deleted. The final outcome is to produce an understandable flow of operations for the proposed system. This is shown in **Figure 13**.

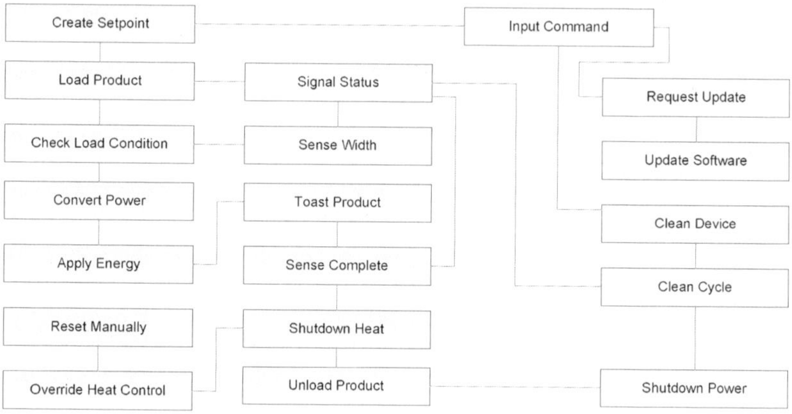

Figure 13: Simplified Functional Analysis

Once the Simplified Functional Analysis has been diagrammed to the satisfaction of the project planner, these functions may be grouped into similar functional families, which will ultimately be used to define the major subcomponents necessary to successfully accomplish the project. This is shown in **Figure 14**.

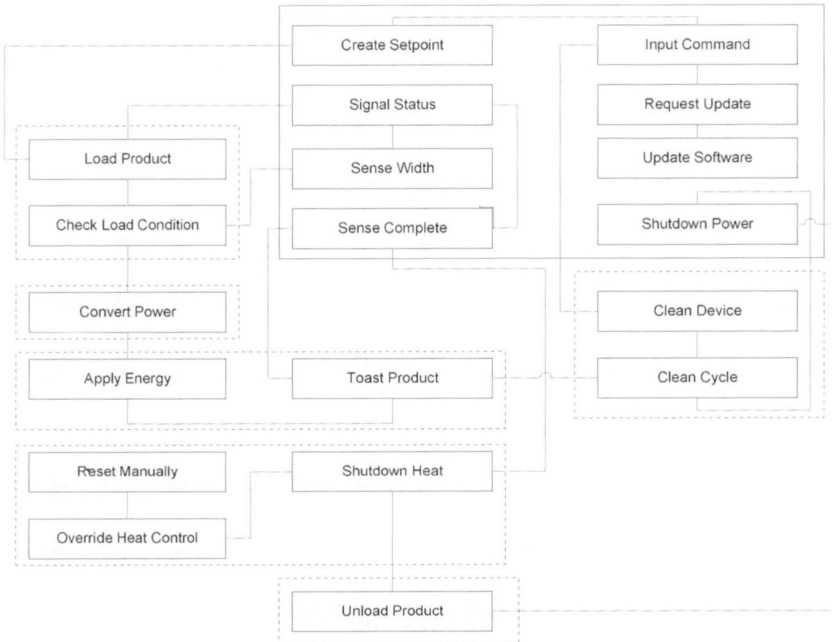

Figure 14: Grouped Functional Analysis

Once the Functional Analysis has been grouped, these groups can be collapsed into collectives or subcomponents that must be provided by the project. This is of particular importance as it defines the major subprojects that will be implemented during the execution of the project, and will coincide with the major headings in our hierarchical task listing. As can be seen, we have identified the top level of our system, the first level of indentation and the second level of functions that must be generated in order to complete the project. This construct will be extremely useful in the development of our work breakdown structure (WBS), which is the list of hierarchical tasks needed for the completion of the project. The WBS is also the most significant input to our schedule. This is shown in **Figure 15**.

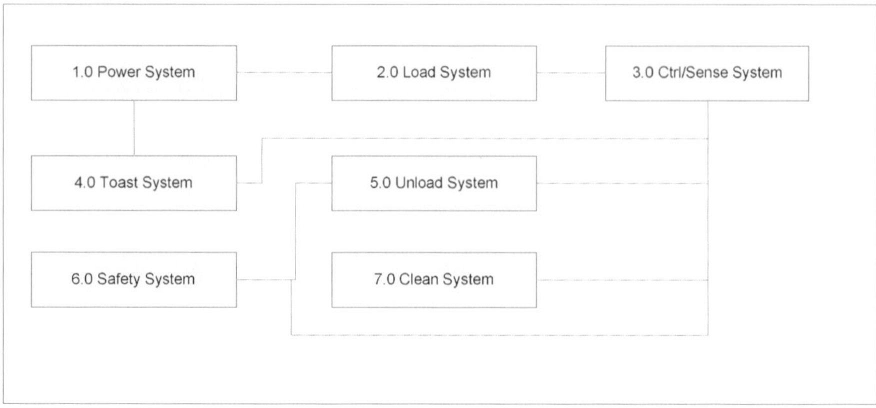

Figure 15: Subcomponents of Toaster

Once this effort has been completed, we are ready to begin the development of our hierarchical task list. This list will be utilized as an input to our schedule to define all necessary tasks that must be accomplished on the project. To clarify for our toaster project:

Toaster project
- Power system
- Load system
- Controls and sensing system
- Toasting system
- Unload system
- Safety system
- Cleaning system

To begin the generation of our work breakdown structure, we must translate these needs into tasks:

Toaster project
- Develop power system
- Develop load system
- Develop controls and sensing system

Develop toasting system
Develop unload system
Develop safety system
Develop cleaning system
Integrate systems
Prototype toaster
Test toaster
Release of System

Note that additional tasks have been added to the list to accomplish the integration and testing of the system being developed. The next step is to add additional details to each of the subprojects in order to describe all those tasks necessary to accomplish that milestone. For example:

Develop power system
Layout power system
Generate electrical drawings
Generate wiring diagram
Generate installation drawings
Generate bill of materials
Create documentation package
Release power system

Additional detail can be added to each of these indented task groupings to create a completed hierarchical task list. Detail is typically added to the task list until the final task descriptions become self-explanatory to the reader. It may require many levels of indentation, but the further we delve into the project schedule, the more robust our project plan becomes.

We can see this in an example of making a chocolate cake:

- Chocolate cake project
  - Planning
    - Locate recipes
      - Select chocolate cake recipe
        - Select Grandma's recipe
      - Select frosting recipe
        - Select Four Seasons' glaze
  - Logistics
    - Reserve kitchen for Saturday
    - Clear calendar for Saturday

The same technique is utilized in order to establish the entire hierarchical list of tasks. This is also known as the work breakdown structure, or WBS. Once this list of tasks is complete, we can incorporate it into a variety of project planning software packages. This WBS is inserted into the task description column, as shown in **Figure 16**. This format of project plan is known as the Gantt chart, and is characterized by a hierarchical list of tasks associated with a calendar.

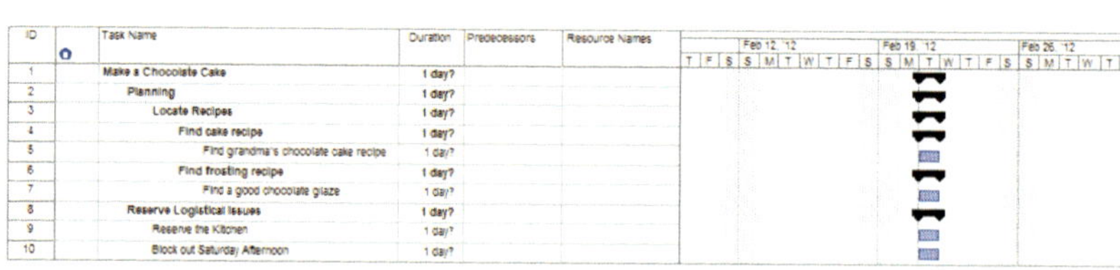

Figure 16: WBS input to Gantt Chart

Note that on the Gantt chart, there are ten independent tasks. Of these ten tasks, only four are shown as bars in the diagram. These are also coincidentally the tasks that are indented the furthest to the right throughout the project plan. As it turns out, these are the only tasks that require work to be completed. All other levels of tasks that have been out-dented from these tasks are simply summary tasks and are used as placeholders in the project plan (they define the hierarchy.)

This means that we have only four tasks or work packages that must be accomplished to complete this scheduled project. Software packages such as Microsoft project and ProjectLibre take care of this identification automatically for us.

Our next challenge is to identify the relationship between the various tasks on the project plan. This is normally accomplished by linking the tasks one to another in order to clarify the order in which these tasks should be accomplished. The most common method links the finish of a task to the start of the following task. Other methods such as the start of a task may be linked to the start of another task as well. Ultimately, this linking enables us to establish the total duration of the project. This is shown in **Figure 17**.

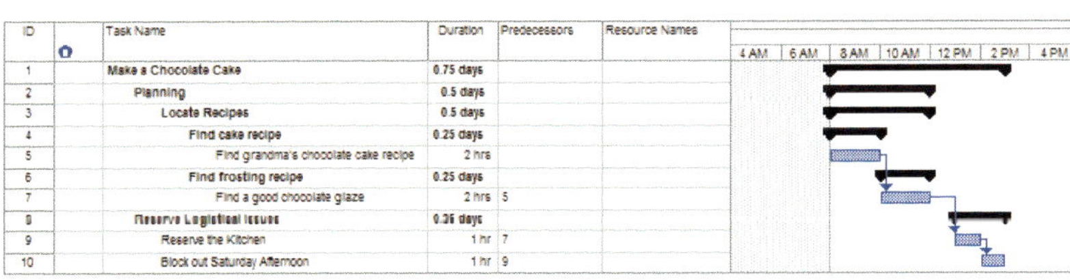

Figure 17: Linked Project Schedule

Durations for each of the work package tasks must be defined at this point. These are input into the Gantt chart, and result in a chronological view of the project to be completed. Only the work package tasks are assigned durations and linked. All other tasks are summary tasks and simply follow the positioning of the work packages.

Durations are estimated based upon the planner's knowledge of the work to be completed. In cases where the planner does not have sufficient knowledge to feel comfortable with these estimates, the aid of a subject matter expert is used. In some cases, if completed project plans are available with actuals that are similar to the tasks being planned, those actuals might be used as a basis of estimate.

At this point, resources are assigned to the work packages in order to establish what skill areas (and therefore what billing rates) are necessary to accomplish those tasks. This is shown in **Figure 18**.

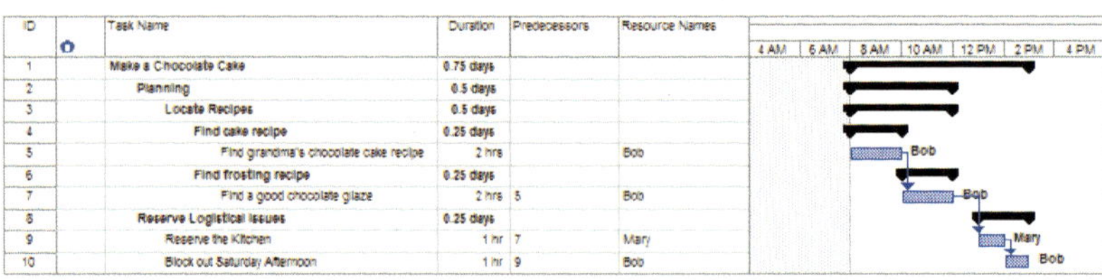

Figure 18: Resource Assignments

Resources are only assigned to the work packages. This ensures that resources are only paid for once in the project's final roll up of the budget. If resources are added to the summary tasks, the project leader will pay for those resources multiple times during the execution of the project.

Indentation levels of the task descriptions are related to the hierarchical task construction, whereas the indentations on the schedule portion of the Gantt chart are chronological in nature. These levels of indentation are entirely independent of each other.

Utilizing this methodology, we have successfully generated a detailed project plan to make a chocolate cake. There are multiple methods that can be used to accomplish this, and any means by which the project planner can generate an effective work breakdown structure and Gantt chart are considered reasonable in the generation of the project plan.

Networking Method

A second method of plan generation is the networking method. Imagine you have been provided a list of tasks to be completed, in no particular order. Begin by assigning a letter to each of the tasks:

(A) Bake ham
(B) Make sauce
(C) Warm buns
(D) Cut ham
(E) Preheat oven
(F) Buy food
(G) Get recipes
(H) Make salad

Build a network of the tasks based upon their interrelationships with each other. Then assign a duration value between the tasks representative of the effort required for completion. This is an event-based system, and allows easy modification of the network as tasks are added or deleted. This network is shown in **Figure 19**.

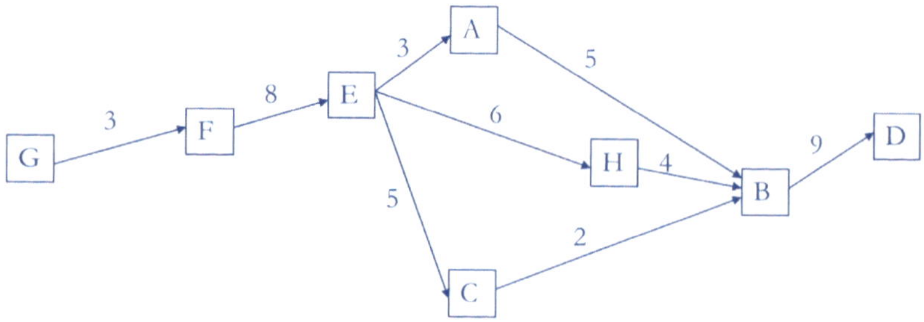

Figure 19: Network Diagram

The network diagram has a significant advantage in that it can help us to identify the critical path. This critical path is of significant importance to us as it represents the longest path from the start of the project to the end of the project. Those tasks that fall along the critical path are critical due to the fact that any delay in the critical path results in a commensurate delay in the delivery of the project. The critical path in this case is shown in **Figure 20**, and is comprised of tasks G-F-E-H-B-D.

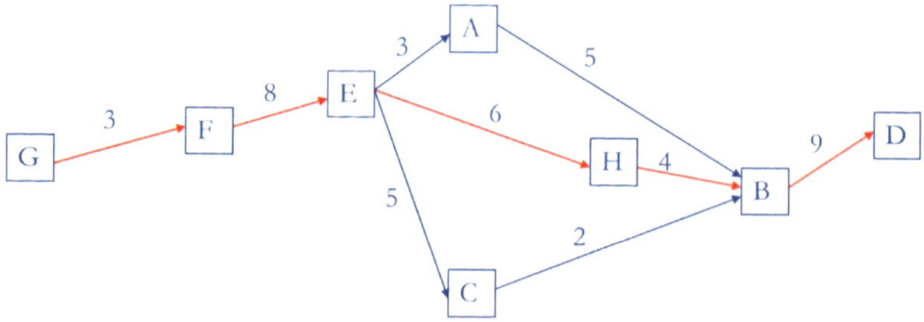

Figure 20: Network with Critical Path

In this instance the critical path contains ten units of time between tasks E and task B. The alternative paths require eight and seven units respectively. This means there are several units of float associated with each of these pads, which actually means that the eight units of work may float anywhere within the ten units allotted and still finish without impacting the completion of the project.

The network method of scheduling is often used by project planners due to the perception that it requires less work than the production of the Gantt chart. However, these two approaches contain the same information for the project plan. In fact, if you choose to transition from a Gantt chart to network chart, the methodology is quite straightforward. First we will begin with the Gantt chart, as shown in **Figure 21**.

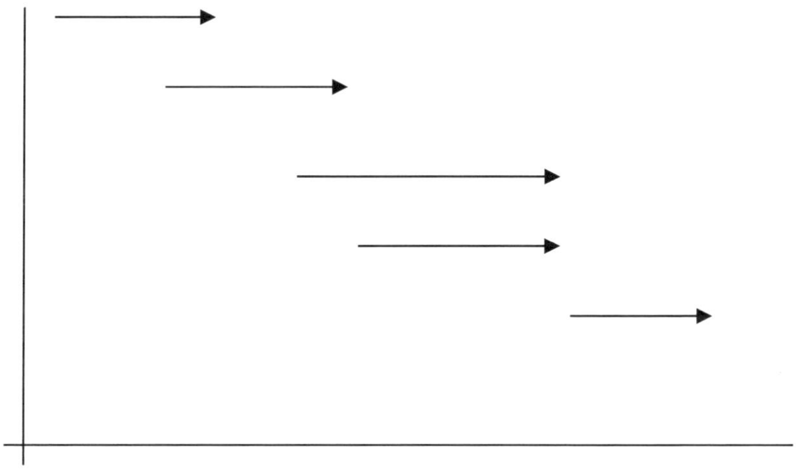

Figure 21: Gantt Chart

Next we will assign an event to the start and finish of every task, seen in **Figure 22**.

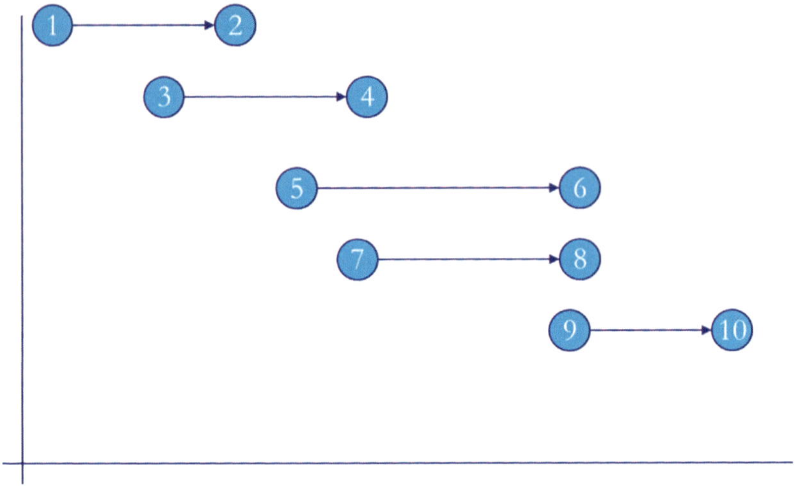

Figure 22: Events assigned to Gantt

We then assign relationships between the events representative of the relationships between the tasks on the Gantt chart, seen in **Figure 23**.

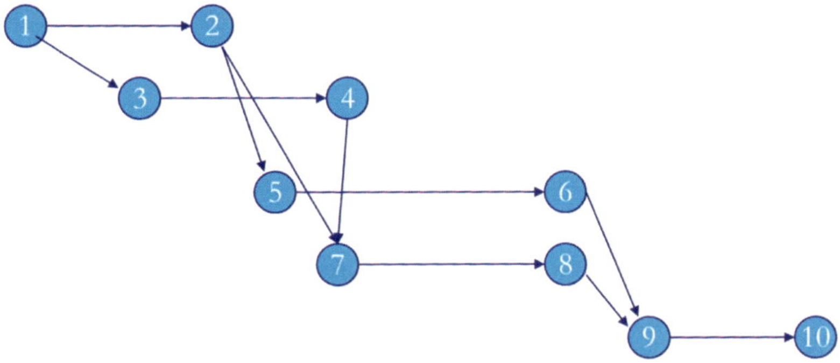

Figure 23: Network Diagram

Durations are added to the links and the critical path is identified, seen in **Figure 24**.

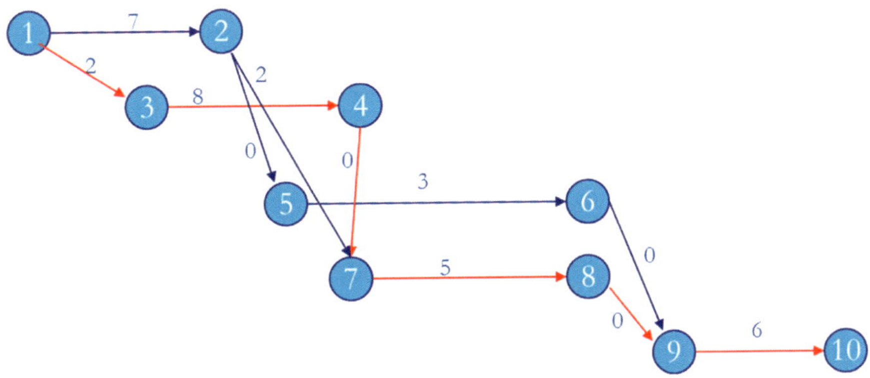

Figure 24: Network Diagram with Critical Path

The Preferred Method of Schedule Generation

The method that is utilized in many large companies for the generation of an appropriate schedule for a project plan includes the development of an Integrated Master Plan (IMP) and an Integrated Master Schedule (IMS). This method is relatively straightforward.

There are three steps to completing the integrated master plan.

1) Identify events
2) Identify accomplishments
3) Identify criteria

To complete an integrated master schedule, expand the task hierarchy to include all of the work packages necessary to accomplish the criteria, accomplishments and events. A simple software project example is included in **Figure 25**.

IMP Level	IMP ID	Task Name
Event	5.0	− Internal Integration & Pre-Qual Test Complete
Accomplishment	5.1	− GNC Software Builds Complete
Criteria	5.1.1	+ GNC Software Build 1 Complete
Criteria	5.1.2	+ GNC Software Build 2 Complete
Criteria	5.1.3	+ GNC Software Build 3 Complete

Figure 25: Integrated Master Plan

The addition of tasks will complete our integrated master schedule in much the same way that we developed our Gantt chart. The completed integrated master schedule is shown in **Figure 26**.

IMP Level	IMP ID	Task Name	CWBS	Duration	Early Start	Early Finish
Event	5.0	− Internal Integration & Pre-Qual Test Completed		189 d	9/1/09	5/21/10
Accomplishment	5.1	− GNC Software Builds Completed	1.1.6.5	189 d	9/1/09	5/21/10
Criteria	5.1.2	− GNC Software Build 1 Completed	1.1.6.5a	63 d	9/1/09	11/26/09
Task	5.1.2	Determine GNC SW Build 1 Requirements	1.1.6.5a	10 d	9/1/09	9/14/09
Task	5.1.2	Conduct GNC SW Build 1 Design	1.1.6.5a	18 d	9/15/09	10/8/09
Task	5.1.2	Code GNC SW Build 1	1.1.6.5a	25 d	10/9/09	11/12/09
Task	5.1.2	Conduct GNC SW Build 1 Unit Test	1.1.6.5a	10 d	11/13/09	11/26/09
Criteria	5.1.3	− GNC Software Build 2 Completed	1.1.6.5b	63 d	11/27/09	2/23/10
Task	5.1.3	Determine GNC SW Build 2 Requirements	1.1.6.5b	10 d	11/27/09	12/10/09
Task	5.1.3	Conduct GNC SW Build 2 Design	1.1.6.5b	18 d	12/11/09	1/5/10
Task	5.1.3	Code GNC SW Build 2	1.1.6.5b	25 d	1/6/10	2/9/10
Task	5.1.3	Conduct GNC SW Build 2 Unit Test	1.1.6.5b	10 d	2/10/10	2/23/10
Criteria	5.1.4	− GNC Software Build 3 Completed	1.1.6.5c	63 d	2/24/10	5/21/10
Task	5.1.4	Determine GNC SW Build 3 Requirements	1.1.6.5c	10 d	2/24/10	3/9/10
Task	5.1.4	Conduct GNC SW Build 3 Design	1.1.6.5c	18 d	3/10/10	4/2/10
Task	5.1.4	Code GNC SW Build 3	1.1.6.5c	25 d	4/5/10	5/7/10
Task	5.1.4	Conduct GNC SW Build 3 Unit Test	1.1.6.5c	10 d	5/10/10	5/21/10

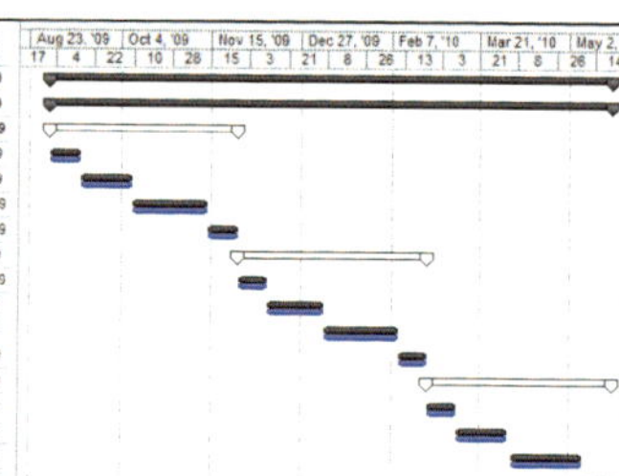

Figure 26: Integrated Master Schedule

This method is very straightforward and develops a highly detailed schedule that provides a coherent view of the project to be performed. As a final example in this section, we will provide the definition of a project schedule for a holiday dinner.

Project name: Holiday Dinner
Requirements:
Ten guests
Turkey with stuffing
Side dishes: potatoes, carrots, rolls, and cranberries
Pre-meal wine
Dessert: apple and pumpkin pie

To define Events:
Holiday Meal

To define Accomplishments:
Holiday meal
    Invite guests
    Prepare for meal
    Prepare meal
    Serve meal
    Serve dessert
    Reset home

To define Criteria:
Holiday meal
Invite guests
    Prepare for meal
        Clean house
        Prepare kitchen
Prepare dessert
    Prepare meal
        Prepare turkey
        Prepare side dishes
        Set dining table
    Serve meal
        Serve wine
        Serve dinner

Clear dinner

Serve dessert

Serve dessert

Clear dinner

Reset home

Clean kitchen

Clean remainder of home that has been used

We then detail the task structure, creating the IMS (Integrated Master Schedule), shown in **Figure 27**.

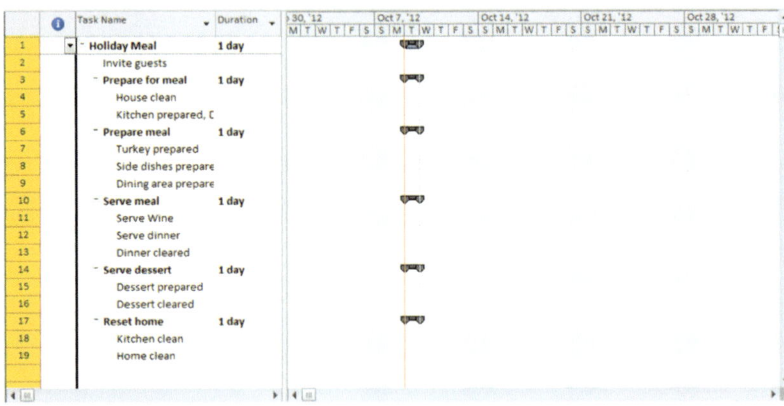

Figure 27: IMS

Add the task durations and link the tasks together:

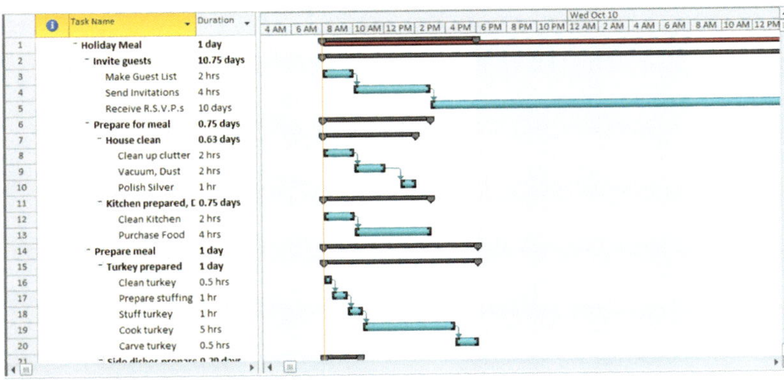

Figure 28: IMS with Links

Link the subprojects together:

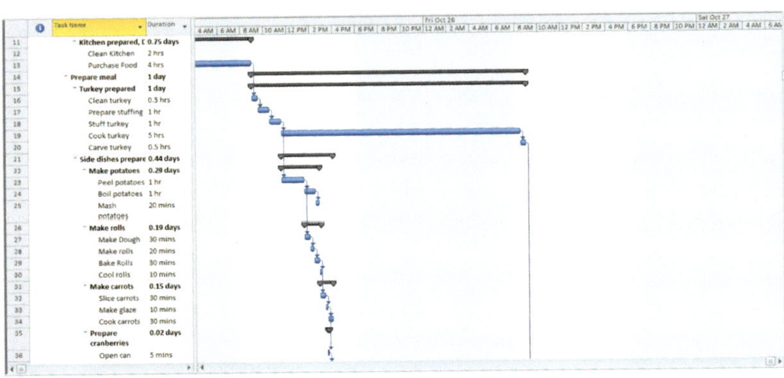

Figure 29: Linked IMS

Then assign resources and complete your project schedule.

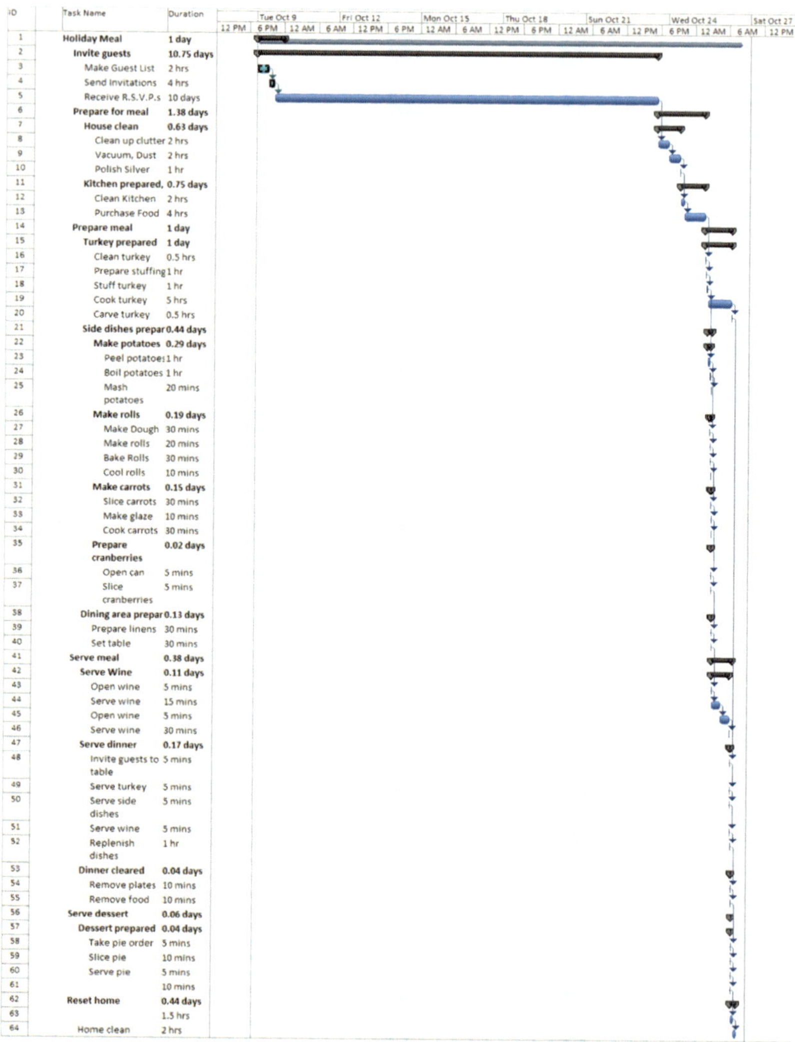

Figure 30: Completed Project Schedule (IMS)

Evaluating Schedule Risk

Risk assessment is critical to the success of the project manager. It is common for the average project manager to begin the plan with a high likelihood of failure. This is largely due to the planning of a project irrespective of other work occurring within the organization. Risk assessment allows us to define the potential challenges that will arise during the project execution phase and account for them in our planning.

To define risk, consider each task as a separate project. We can postulate that a given task repeated by a wide variety of people will result in a variety of outcomes. In this case, we are interested in the completion time of the task. We see by the following diagram that the resulting outcome of the task can be modeled by a normal distribution.

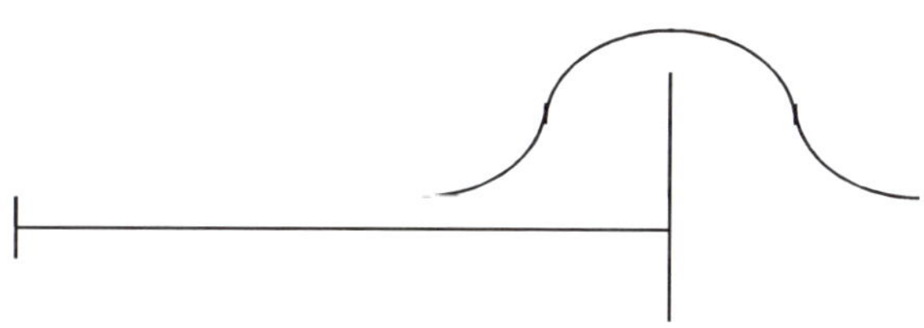

Figure 31: Task Completion Distribution

The issue of risk assessment is then related to defining the approximate shape of the distribution about the nominal task end. This shape will define for us the likelihood of completing the task as expected. Taken as a whole, all tasks within a project can be assessed in this manner, allowing a confidence interval to be calculated for the entirety of a project.

The information to define this distribution can be collected in a straightforward manner as shown in the following table.

Task #	Nominal Duration	Worst Case Duration
1	4d	5d
2	1d	1d
3	5d	40d
4	6w	7w
5	10d	11d

As expected, the nominal duration is the intended schedule time to complete the work. The worst-case duration is defined to assess the approximate width of the standard deviation. It is surprising to observe that several of the tasks show a distribution width that exceeds the task length. This is an insight into the psyche of the project planner, as he or she does not desire to communicate that the task schedule is thirty-five days plus or minus one year. This creates a significant credibility issue for the planner, and is therefore mentally suppressed.

We desire to communicate a schedule with realistic risks associated, and therefore a worst-case scenario is chosen as the appropriate method, bypassing this blind spot in the planner's viewpoint. As we observe from the above table, the risk of certain tasks is quite low and predictable. Other tasks show a great deal of risk, such as task three. These of course are the tasks requiring the attention of the project manager, as the greatest variability in response will be observed in this area.

For example, consider a project made up of the following tasks:

Task #	Nominal Duration	Worst-Case Duration
1	10d	15d
2	10d	20d
3	10d	10d
4	10d	13d

The project at the outset shows expected times of completion:

$$T_{nominal} = 10d + 10d + 10d + 10d = 40\ d$$

$$T_{worst\ case} = 15d + 20d + 10d + 13d = 58\ d$$

This provides us with an idea of the standard deviation, assuming the worst case shows a 3-sigma distribution.

$$\sigma = 6\ d$$

Therefore, if we are asked if we are confident that we will complete the project in forty-six days we statistically predict that this encompasses a nominal value of 40 d plus a one-sigma area of the distribution. This corresponds to a confidence interval as follows:

$$C.I._{46days} = 67\%$$

In the event twenty-six days have passed (about one-half the management-allowed time) and the first two tasks are completed, we can recalculate the confidence interval.

$$T_{nominal} = 10d + 10d = 20\ d$$

$$T_{worst\ case} = 10d + 13d = 23\ d$$

$$\sigma = 1\ d$$

We must consider our confidence now of completing the remaining tasks within twenty-two days:

$$C.I._{22days} = 20\ \text{days} + 2\ \text{sigma} = 93\%$$

Therefore, although we are approximately one-half completed with the project, and one-half completed with the tasks, our confidence interval has increased due to the lower risk tasks still remaining to be completed. If we are able to shift the higher risk tasks further forward into the project, or if we are able to focus directly upon these tasks, we can have the greatest impact on the overall project delivery.

Imagine that in the above example, the tasks are identical but the first two tasks are completed within twenty-five days.

Task #	Nominal Duration	Worst Case Duration
1	10d	15d
2	10d	20d
3	10d	10d
4	10d	13d

$$T_{nominal} = 10d + 10d = 20\ d$$

$$T_{worst\ case} = 10d + 13d = 23\ d$$

$$\sigma = 1\ d$$

We must consider our confidence now of completing the task within twenty-three days:

$$C.I._{23days} = 20\ days + 3\ sigma = 99.7\%$$

Therefore, it is very important for the project manager to mix the risks within the project appropriately to provide for lower risks as the project proceeds. Additionally, it should be noted that based on this model, all projects are not including risk calculations are currently planned with a confidence interval of only fifty percent.

Critical Path Method

The critical path method is the most traditional manner in which risk is calculated in project plans. This was developed during the initial stages of Project Management, and is used extensively in most project-oriented organizations. There is, however, a concern about this technology as it is typically utilized within organizations.

As has been previously discussed, it is common to calculate inappropriate risk for any given project task or subproject. In the critical path method, there are three estimates used for this calculation, which include:

Optimistic estimate
Nominal estimate
Pessimistic estimate

These three estimates are then combined with a simple weighted average to establish the estimated risk for that some project or task. This arrangement is shown in Figure 32.

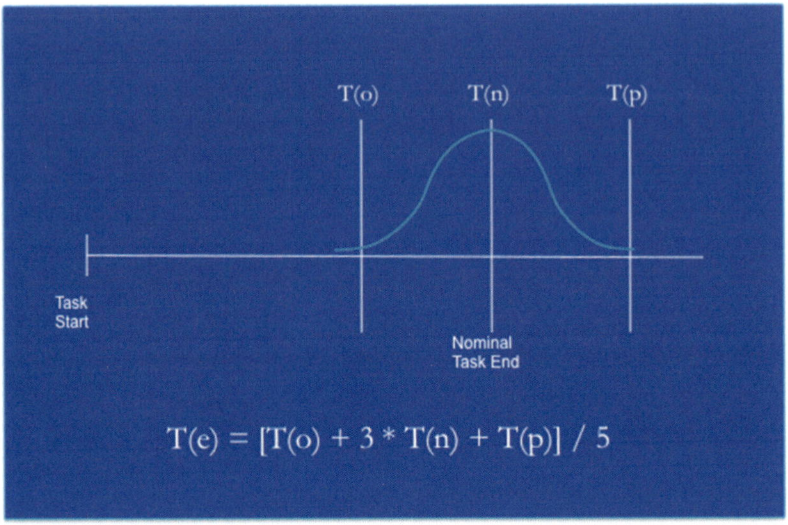

Figure 32: Critical Path Method

The major problem that is discovered when the critical path method is utilized is that the humans involved in the estimation process often circumvent its intention. As can be seen in Figure 33, it is the natural tendency of the project manager to make estimates of the optimistic to nominal interval equal to the nominal to pessimistic interval. As should be obvious to the reader, this will essentially remove all risk from the calculation. The major stumbling block for most project managers is the fact that when this calculation is carried out they see that the final estimated time is equivalent to their original nominal estimates, thereby reinforcing that they have estimated the test correctly at the outset.

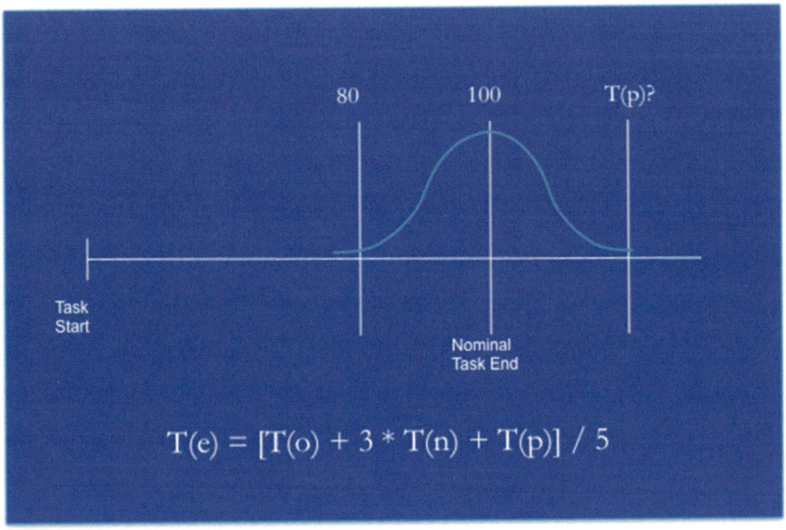

Figure 33: CPM Intervals

It should also be obvious that the previously discussed method of risk calculation is equivalent to the critical path method, but does not allow the optimistic assessment to bias the resulting calculation of risk. Therefore, it is recommended that the project leader utilize the recommended risk method in order to ensure that the resulting schedule has a high probability of success.

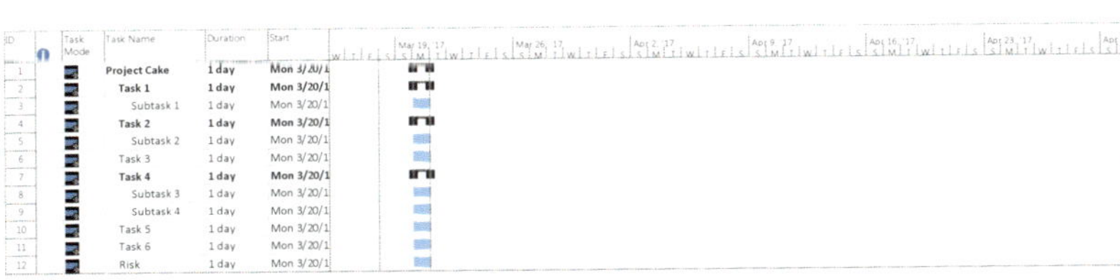

Figure 34: Project Cake planning stage 1

Once the magnitude of the risk has been established, it must be incorporated into the project plan in a manner that allows clear communication of the risk, but does not provide motivation to the project team to utilize all the risk at their disposal. Remember that if the team thinks they have four weeks to complete a task, it will take at least four weeks to do so. Therefore, when incorporating risk we must consider the message being sent to the project team, ensuring that they are held to appropriate expectations in the completion of their work. To accomplish this, we will begin with a project schedule as shown in Figure 34. In this view we have defined seven work packages for this project, and added risk at the end of the project. Note that the total duration of work that is expected to be completed is represented by the work packages, in this case nineteen days of effort. Some of these tasks are completed concurrently, so the total effort should take just fifteen days on the calendar. We would expect to pay for nineteen days of effort in our budget.

Note that the schedule shows an expected completion of twenty days when risk is incorporated. This demonstrates the use of float in project management. What is being communicated is that we expect these nineteen days of effort to be completed within a twenty-day window, and the work can float within that window. Since some of the work can be completed concurrently, we expect that there are five days of float, allowing the tasks to move about within the window while still meeting the end date.

Why is this appropriate as a descriptor of our project? Imagine the typical employee working on your project. The phone rings, and he is asked and answers several questions about his previous project in a matter of minutes. Technically, those minutes should be charged to the previous project. In reality, they will likely come out of the new project budget. Regardless of the account that is charged, the project is delayed. Risk is incorporated into the project plan to adjust the schedule for unforeseen interruptions and issues on your project, which are known to occur. The importance of the end date is not lost, as the work has a high probability of being completed as planned, not necessarily on the date shown on the schedule, but at some time during the allotted window.

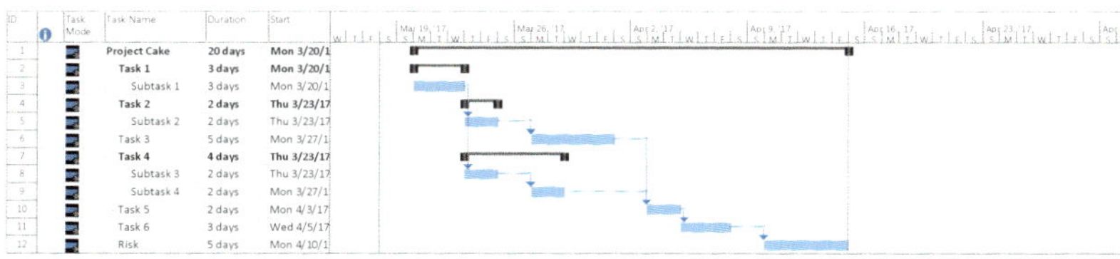

Figure 35: Project Cake Planning Stage 2

In Figure 35, durations estimated for each of the work packages and interdependencies have been defined. Note that the durations are only entered for work packages and not summary tasks, and all interdependencies are defined at the work package level. It is a common mistake for project planners to interconnect summary tasks; however, these will convolute the project in the event major modifications are required.

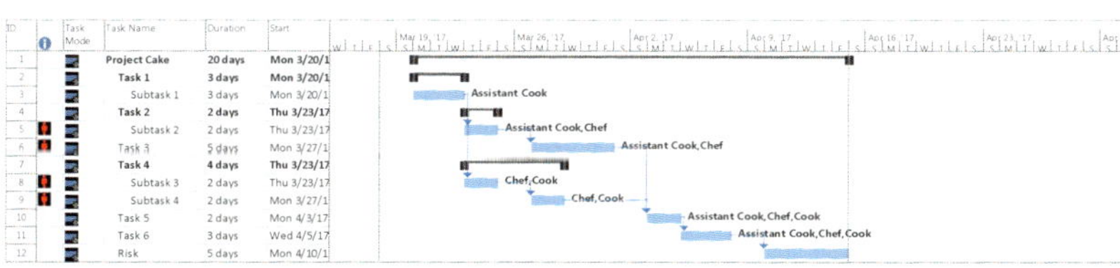

Figure 36: Project Cake Planning Stage 3

As can be seen in Figure 36, the project plan has been completed with the addition of resource allocations for each of the work packages. Note also that the resources are only assigned to work packages; they are not assigned to summary tasks or to the risks task. If they are assigned to summary tasks, the budget will reflect a multiple of the actual costs expected to be incurred during the execution of the project. No individual is too quickly allocated to the risk task in order to represent the float without paying for a developmental resource during that task duration. The security benefit of scheduling your task in this way is that no individual will assume that she can utilize that task in order to keep her own tasks on schedule.

It is also important to utilize functional descriptors for the resources being assigned to the greatest tasks, as particular individuals have not yet been allocated to the project. It is important for the project manager to recognize that the individuals to be assigned are not necessarily committed to the project at this time, and yet the budget must be defined. We understand that if we allocate specific functional roles, we can estimate the expected hourly cost of those roles without having particular individuals assigned. In this way, project risk can be attached to every appropriate section of the project.

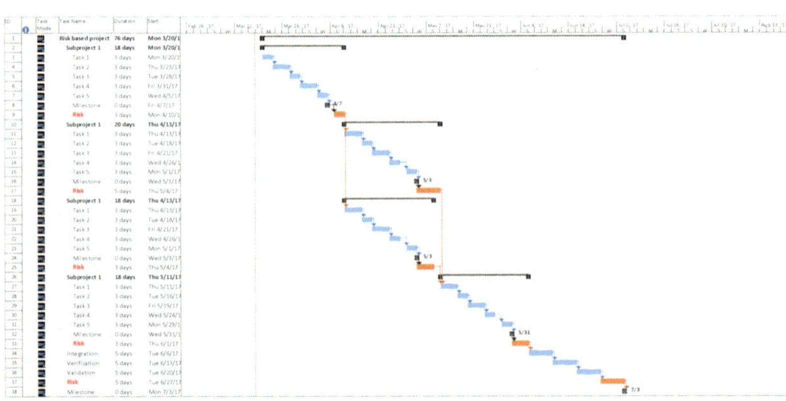

Figure 37: Incorporating risk throughout the project

With regard to the overall project, risk will typically be incorporated in multiple places throughout the project development cycle. As shown in Figure 37, risk is typically incorporated into each of these subprojects. Note that the risk is added after the milestone delineating the delivery of the subproject. This is done in order to communicate effectively to the project team want the expected end date for their work will be. Risk can then be consumed by the project manager as necessary to keep the project on track on a subproject basis.

Note that risk is also incorporated at the very end of the project, and this risk task is utilized to ensure that the project meets the expected end date. In this case the risk is incorporated before the final milestone for the completion of the project. This is necessary to make sure that the expected end date is communicated effectively to those stakeholders external to the project.

The project manager is responsible for the utilization of risk on his or her project, as this is reserve built into the schedule to be used at the project manager's discretion in order to keep the project on track. Should the project effort run ahead of schedule, there is no reason why the risk should be consumed by the project manager, and when risk is not necessary, the remainder of the project can be moved left on the timeline.

Risk Management

While schedule risk is the most commonly addressed by the project manager, there are actually four types of risk with which to be concerned:

Technical risk – Something will not be possible technically
Schedule risk – Our estimates will not be predictive
Budget risk – We will run out of money before we complete the work

Operational risk – If we finish, will the customer want it? Will someone pay us?

Each of these risks must be quantified in a similar manner for us to utilize them as project managers. We therefore translate all types of risk into a monetary value, in order to treat all types of risk concurrently.

Risk: A future event or series of events with a probability of occurrence and the potential for impact to the objectives of the program. This implies that there are two components to assessing risk, the probability and the impact to the program.

The risk management methodology contains a five-step program. These steps include:

Risk identification
Qualitative risk analysis
Quantitative risk analysis
Risk response planning
Risk monitoring and control

Risk identification is one of the most challenging aspects of a risk program. This requires that the system engineer considers all of the complexities of the program and also considers several questions:

What are the things that could potentially go wrong on this program?
What are the possible causes for those things going wrong?
If things do go wrong, what are the potential consequences?
What are the things that could potentially impact the program positively?

These risks are typically described in a narrative form, and must define the specific concern in a manner that will reduce the nebulous nature of the statement in order to provide specificity to the developer. These risks are then logged in a table, which will be utilized throughout the life of the system development.

Qualitative risk assessment is frequently used to ascertain those areas of greatest concern for the system development team. This type of risk analysis will require the assignment of high, medium, or low, to the probability of the risk occurring as well as a high, medium, or low, to the impact on the development project. This methodology is often used for executive summaries of the program, but lacks precision and predictive capabilities.

Quantitative risk assessments require a more stringent analysis on the part of the system developer. This method will require the system developer to ascertain the probability of occurrence by defining a percentage of likelihood with a precision defined by the knowledge of the risk. In addition, the consequences of that risk occurring must be considered and an estimate created regarding the financial impact on the project objectives. These two factors are multiplied to define the potential risk exposure on the project.

Risk response planning includes the potential mitigations that will be implemented regarding the risks identified and quantified. There are several methods commonly used for risk response, which include:

Risk avoidance – Choosing not to undertake the risk

Risk transference – Transferring the risk to another party, such as with insurance
Risk mitigation – Changing the risk in some way, by addressing the probability or impact
Risk deferral – Choosing not to accept that risk at this time
Risk acceptance – Choosing to accept the risk as it is defined

In general, the risks identified will be identified by number and mapped onto what is commonly referred to as a heat map in the risk management process. The qualitative assessment is shown in **Figure 38**.

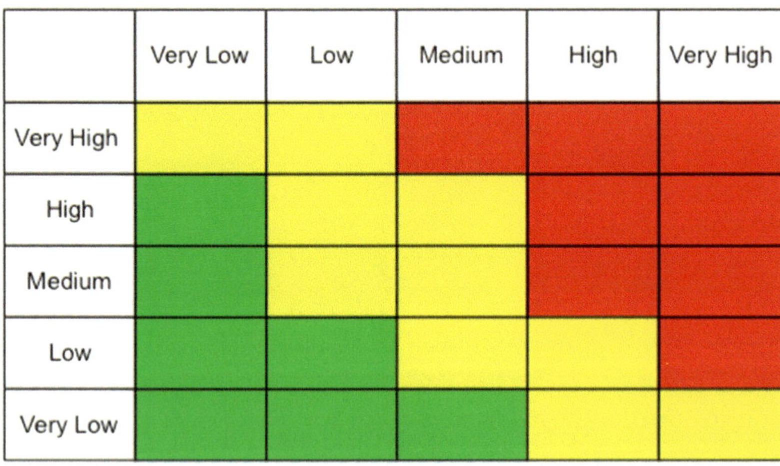

Figure 38: Qualitative Risk Matrix

Quantitative risk management is typically tracked via spreadsheet, where each risk is identified, assigned to a specific point of contact, and the probability and impact are identified, enabling the calculation of an estimated monetary value. This estimated monetary value is then summed for the program, which provides an assessment of the financial exposure existing on the program. An example matrix is shown in **Figure 39**.

	Description	POC	Probability	Impact	EstMonValue	Actions
1	Heating Method doesn't work	T. Jones	0.5	$1,500.00	$750.00	
2	Heat at outer case causes safety issues for people	R. Williams	0.75	$2,000.00	$1,500.00	Add insulation around canister
3	Risk of cord getting into toaster and burning off insulation	R. Williams	0.2	$3,500.00	$700.00	Reduce length of cord to 15 inches
4	May miss schedule due to delivery of heating core delay	M. Sharpe	0.85	$12,000.00	$10,200.00	
5	Customer may not like surface finish	E. Smith	0.05	$100,000.00	$5,000.00	
6	UL Listing test	R. Miller	0.35	$15,000.00	$5,250.00	Use UL Consultant in case of failure
7	Power supply noise below required threshold	F. Meeker	0.8	-$12,000.00	-$9,600.00	Remove filter from design if it works
	TOTAL RISK				$13,800.00	

Figure 39: Risk Matrix / Register

The quantitative risk management requires a greater degree of analysis and insight into the risks identified for the program. Due to this, once the risks have been identified they can be effectively normalized for use in the risk heat map. This is useful in order to establish the relative importance of those risks for management. To accomplish this, one must examine the lowest and highest probabilities identified, and evenly spread those probabilities across the Matrix fields. Secondly, the impacts are assessed for maximum and minimum and these are used as guides to create the impact buckets that will be used for the risk assessment. An example of this is shown in the following quantitative heat map.

Impact

Probability	<20K	<40K	<60K	<80K	<=100K
<=1	4				
<.8	2				
<.6	1				
<.4	6				
<.2	3				5

Figure 40: Risk Quantitative Heat Map

Figure 40 identifies the mapping of the risks identified in the previous risk matrix. In this particular case, attention would be paid to risk number five, as this provides the greatest threat to the success of the system development. In general, those risks identified in the yellow or red region are typically targeted for mitigation strategies in order to reduce the exposure of the system development.

Risk management is the process of monitoring and adjusting these risks on a systematic basis during the program development cycle. In most cases this translates to a review of the risk register during regular project meetings, where each risk is assessed to define if any mitigation strategies have been implemented or if occurrences on the project have changed either the probability or impact of any of the risks. The risk register and the corresponding heat map are then updated and communicated during the regular business rhythm within the organization.

Once we have completed the integrated master schedule, we have completed the second task in our project planning challenge, namely we have defined "how long it will take." Next, we move on to developing the budget.

Developing the Project Budget

To define an appropriate budget for our project, we must first make sure that our Integrated Master Schedule makes sense. To accomplish this, we will meet with the Functional Managers to ensure the schedule we have generated has realistic task identification and durations that will produce a reasonable expectation of completing the project as planned. We meet with Functional Managers because they are the subject matter experts for the areas of interest, and it is very likely they will help us to produce a more accurate and representative schedule. What's more, since the Functional Managers will assign their personnel to this project, this is our first step at socialization of our upcoming project plan.

For each of the work package tasks within our Integrated Master Schedule, we will assign costs as appropriate for that task. The resources that are utilized to accomplish that task and their associated billing rates typically represent these costs. These billing rates are multiplied by the duration of the tasks to generate the actual cost of those resources accomplishing that task.

Indirect labor hours
Fixed costs
Materials
Purchased parts
Subcontracts
Freight
Travel costs
Other miscellaneous costs (10-15% add-on)

These can be incorporated into a simple project budget worksheet, such as included here for reference.

Cost Estimating Worksheet

Project Title:						Project manager:
Activity	Schedule	Budget				
No.: Name/ Description	Hours	Labor hr x rate	Material	Other	Total	Resource Manager

As the budget is completed, we have accomplished what we originally set out to achieve. We have defined:

What will we do?
How long will it take?
How much will it cost?

Our only remaining task is to complete the Project Charter. The charter document is a single document that incorporates all of the work accomplished to date on the project plan.

The Project Charter Document is used in a similar manner as a contract within the organization. The information is filled out at the completion of the project planning process, and defines exactly what will be accomplished on the project, what will not be accomplished on the project, the expected costs and the expected timing of the project. Space is provided to gain approval signatures from stakeholders on the project and the intent of the project charter is to use the document both to open and close the project.

An example Project Charter Document is included following some examples to provide some insight into the technical information required. We have already defined all the information required to complete the document, but the charter acts as a single repository for the information.

An Example Project: Dinosaur World

To illustrate the appropriate generation of the detailed specification, we are going to consider a fictional project entitled "Dinosaur World." We have chosen to begin the development of a new theme park that will provide unique and exciting experiences for our customers.

Of course, if we introduce this concept to a variety of our project team members, one of the problems we face immediately is that each of the members of the team currently has a very different idea of what the end product will be. Part of the reason we generate a detailed specification is to ensure that everyone who is associated with this project has a similar or identical understanding of what will be accomplished on the project.

We will begin by defining a few basic characteristics for our project.

Title: Dinosaur World project

Goal:
Produce the Dinosaur World theme park and become a profitable company

Strategy:
Innovation

Statement of work:
We will develop the only theme park that will contain real dinosaurs as well as scientific education.

Tactics:

Staffing
Location
Buildings
Tours
Flora and fauna
Safety

Users:

Patrons
Scientists
Staff
Vendors
Landowners
Government
Insurance
Flora and fauna
Transport
Investors
Project team
Energy

User needs:

Experience
Safety
Certification
Service
Accuracy in offering
Good scientific data
Opportunity to do research
Funding

Investment
Job experience
Opportunity to advance
Working conditions
Housing
Food
Business access
Contracts
Taxes
Tourism
Compliance
Promotional
Low-risk
Natural state
Return on invested capital
Specification, schedule, budget
Quality
Success
Infrastructure
Safety (yes, it is that important)

Specification:

Build the park
- Land
- Build buildings
- Make display areas
- Building closures

Conduct Tours

To increase our resolution, we drill down:

Land:

Island location
Permits
Zoning
Government
Not prone to natural disasters
Temperate climates
Easy access
Close to mainland
Freshwater source
Appropriate Flora for food
Energy source
    Geothermal, solar, wind, hydroelectric

Displays:

Materials
Large team and staff– Scientists, staff, electricians, artists, marketing
Location
Arrangement of displays
Accessibility of displays
Weather resistant
Safety and security of displays and patrons
Vendors of supplies
What dinosaurs are we displaying?

Buildings:

Housing building
Information area
Research lab
Rest area– Bathrooms and common areas

Team of architects, Contractors, industrial designers
Define size, structure, plumbing, safety, fire, flood, and dinosaur attack requirements
Material selection
Blend with the environment– Natural, Camouflage
Display for dinosaur artifacts, Museum style building

Enclosures:

Permits
Limit dinosaur type and size
Camouflage enclosures
Maintenance access for cleaning and animal care
Design to reduce material delivery costs
Safety for the dinosaurs
Safety for the patrons
Sequence by time period

Conducted tour:

24 hour monitoring system
Trained personnel and guards
Walking tour – build a walkway access
Individual station– Provide personal experience, special trained to her guidance
Bus tour – schedule 3 times the day
Helicopter tour – any time that the park is open, schedule ahead
Boat and river to her
Multiple day package deals
Design event celebrations – Anniversary, wedding, birthday

As can be seen in the example, the development of a specification is simply adding fidelity to the planned system by expanding upon the detail required. The specification is considered completely defined when everyone on the development team understands exactly "what is intended to be accomplished" during the development of the project. If there are nebulous or vague items that people do not understand, additional detail must be added to the specification in order to avoid future challenges.

Once the specification has been completed, the major deliverables can be utilized to develop the work breakdown structure and schedule for the project. The completed project schedule is included here for reference in Figure 41.

Figure 41: Dinosaur World Schedule

Project managers often find themselves in situations where the information for a project is presented in a slightly unique way, requiring them to approach the problem utilizing slightly modified techniques. As an example of this, imagine a case where the tasks for a project are listed, but they may or may not be relevant to the project and there may be multiple paths by which the work to be completed. This might be the outcome of a group meeting discussing what needs to be done on a complex project, or simply a random collection of thoughts that someone has recorded regarding the given project. In the **Critical Path Challenge**, we work through the development of a project schedule from a completely different direction.

The Critical Path Challenge – Getting the Goods

Scenario:

You have been selected to take part in a reality game show whereby you are tasked with infiltrating a museum in Casablanca. A painting has been placed on display in the museum that you must safely secure and remove from the building. There are a series of tasks that must be completed by your team, all of whom will work together in order to complete your mission.

You must decide what tasks you will complete, who will complete them, and in what order they will be done. The museum has a simple layout, which makes the job a bit easier. The entrance is in Room C, the alarm is in Room D, and the Painting is in Room A. The alarm will sound ten minutes after passing through an interior door in the museum, or if anything is disturbed in the museum.

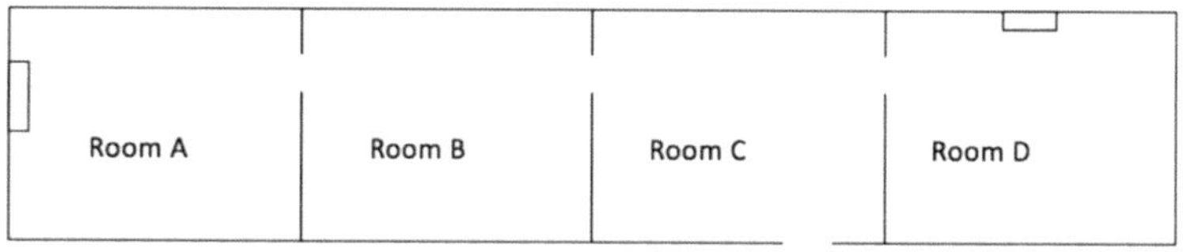

The guard makes rounds every thirty-seven minutes around the exterior of the museum, and you can rely on his sticking to the schedule. Can you get the painting out of the museum and get your team to safety? The potential tasks are listed in the following table.

	Task	Time Required (minutes)
1	Enter Museum	2
2	Exit museum	2
3	Disarm Alarm system permanently	15
4	Package painting for safe transport	12
5	Travel from room to room (from C to B, or across A, C or D)	1
6	Place fake painting on wall	4
7	Open interior door	2
8	Re-arm alarm	5
9	Escape from area of museum	3
10	Close exterior door	1
11	Carry painting from room to room (Same as 5 with painting)	2
12	Check on guard (Identify guard's position)	3
13	Create diversion	1
14	Break open exterior door	6
15	Delay Alarm system for 10 minutes	7
16	Hide painting in museum	5
17	Remove painting from wall	5
18	Close interior door	1
19	Pick lock on exterior door	3

To address this problem, we must consider which tasks are relevant for our effort. In order to fully understand this, we must go through our standard methodology of developing the goal and strategy of our project, which should enable us to begin the development of an appropriate work breakdown structure. From this, we should be able to define a practical schedule of tasks, which will accomplish all of the requirements for the project as well as answer the question as to whether it is possible to get our team to safety.

Museum Heist work breakdown structure

Goal: To possess the painting

Strategy: Get into the museum and out of the museum with the painting without being caught

We can construct this utilizing the methodology:

Events
Accomplishments
Criteria

- Possess painting
  - Get painting
    - Enter the museum
    - Travel to the painting
    - Take the painting
    - Exit the museum
  - Get away
    - Enter without notice
    - Disable alarm as necessary
    - Exit without notice
    - Depart without notice

This construction enables us to understand what tasks need to be completed in order to accomplish the goals set out in the project requirements. Obviously, there is additional complexity that must be incorporated, as the two major accomplishments of getting the painting and getting away are actually intertwined throughout the development of this project. Therefore, we will address each of them separately at first, assuring that we can complete these independent tasks appropriately, and then examine how the two parallel paths must relate to one another.

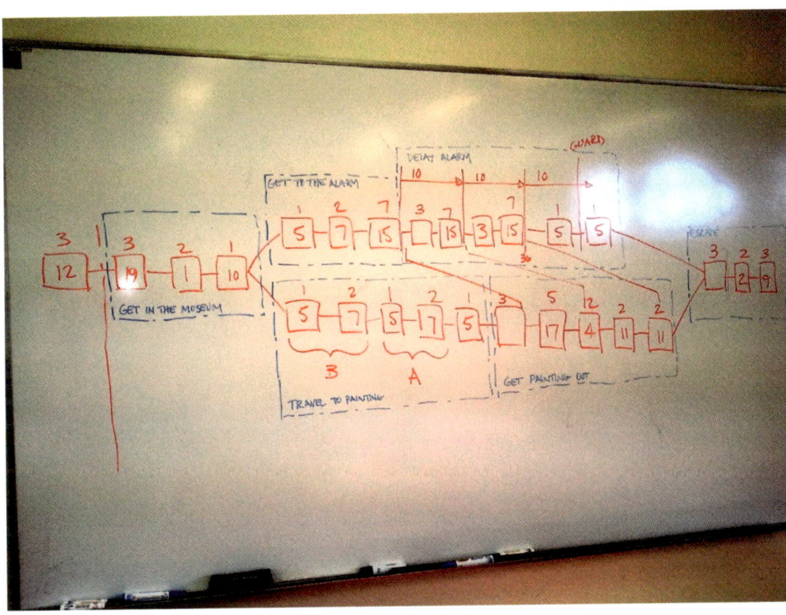

Figure 42: Critical Path Challenge approach

As can be seen in Figure 42, several subprojects have been developed in order to accomplish the two major subprojects for this challenge. After completion of the individual paths, some effort is made in order to ashore and understanding of how the pads are synchronized within the system. This will result in the insight that we must in fact disable the alarm several times during the process of getting the painting. Additionally, we must consider the overall direction of our project as it relates to the position of the guard outside. This complexity can also be readily managed by looking at the parallel paths in any time-based diagram as shown in the Gantt chart in Figure 43.

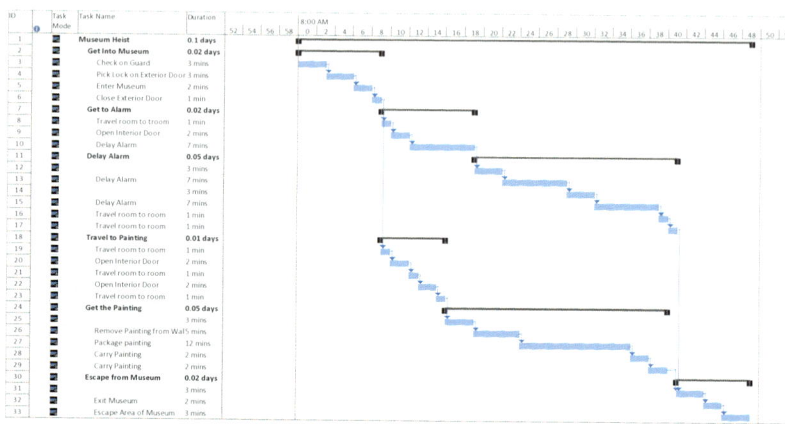

Figure 43: Critical Path Gantt Chart

As can be seen in this example, we can arrive at an appropriate specification and schedule from a variety of approaches. It is critical that we understand the variables presented to us, and address the need for an appropriate project plan based upon this data. In cases where we have incomplete data available, we must devise methods by which we can either estimate or define those variables necessary to complete an appropriate project plan.

The Finley Project

The Finley project has been provided as an additional example of how a detailed project plan can be developed based upon the limited information typically provided to the project manager at the outset of the project. Remember, it is often the case that there is far more information available than may seem readily apparent. We must take some time to clarify our own understanding of the constraints and requirements of the project, and draw some practical conclusions related to the areas that are less defined at the outset of the project. This project is a good example of how the ability to draw some conclusions enables us to complete a practical project plan.

Background

The project is focused on providing a box lunch service delivered to a customer's home on a weekly basis, in order for the family members to have nutritious midday meals during the week.

The Customer

We know the following:
Designed for families
Probably not concerned a great deal about money
More interested in consistency in meal quality
More interested in saving time

So what does this tell us about the customer?

Strategy

We know we are NOT in the business of:
Cost (customer does not care about this)
Flexibility (we do not want to provide many options)
Dependability (if they want superb food, they will go to a restaurant)

Therefore, we are likely in the business of:
Quality (serve a good lunch, every time)
Innovation (this is a unique service model)

Understanding our project alignment, we can now start the project planning.

Goal:
Grow our business with this new market

Values statement:
We provide lunches in a turnkey format to save our customers valuable time without compromising health

Mission statement:
We will produce quality, nutritious meals, delivered reliably to our customer's door on a weekly basis

Business statement:
By consistently saving our customer's time, we will build a strong recurring business base for our future

Goal-Strategy-Tactics-Tasks

Goal:
Grow our business with this new market

Strategy:
We will focus on quality in our products

Tactics:
Build a lunch architecture (Sandwich, Vegetable, Fruit, Drink)
Develop a variety of options to build with
Develop a lunch preparation system
Develop a packaging solution
Develop a delivery solution
(Note that these become our subprojects later)

Define our Users:
Workers
Office managers
Parents
Schools
Day Care providers
Etc.

Needs:
Nutrition
Safety
Time savings
Convenience
Portability
Consistency
Variety

Specification:
5 lunches per week
Delivery weekly to single address
Meal: Sandwich, Vegetable, Fruit, and Drink
Package: fit into 10X4X5 inch space

Designed for refrigerated storage up to 6 hours prior to consumption
Nutritional needs met for age group
Different menus rotating on a 90-day cycle
Packaged in bio-degradable packaging
Etc.

Schedule:
Events
Lunch Meal Delivery

Accomplishments
Develop system to create lunches
Execute system to deliver lunches repetitively

Criteria
Lunch architecture (Sandwich, Vegetable, Fruit, Drink)
Lunch menu options
Lunch preparation system
Packaging solution
Delivery solution
Execute lunch orders

After completion of this level of aggregation of our project approach, we have completed an appropriate work breakdown structure and hierarchy for our technical approach. This provides us with enough understanding of what must be accomplished on this project and how it will be accomplished that we can move into the development of the detailed schedule.

ID	Task Mode	Task Name	Duration
1		**Lunch Meal Delivery Project**	**114 days**
2		**Develop system to create lunches**	**83 days**
3		**Lunch architecture (Sandwich, Vegetable, Fruit, Drink)**	**24 days**
4		Define criteria for lunch nutrition	1 day
5		Develop sandwich design	2 days
6		Develop vegetable/dip design	3 days
7		Develop fruit design	6 days
8		Develop drink approach	5 days
9		Establish criteria for mixing comp	4 days
10		Define lunch architecture	3 days
11		**Lunch menu options**	**21 days**
12		Develop 30 sandwich recipes	2 days
13		Develop 20 vegetable recipes	3 days
14		Develop 15 fruit recipes	4 days
15		Develop 10 drink options	2 days
16		Establish menu mixing criteria	3 days
17		Check nutrition of all combination	4 days
18		Create 90 day menu	3 days
19		**Lunch preparation system**	**38 days**
20		Establish Kitchen needs	3 days
21		Establish kitchen layout	4 days
22		Define staffing needs	5 days
23		Purchase components	6 days
24		Rent kitchen space	2 days
25		Install kitchen components	3 days
26		Test kitchen workability	4 days
27		Define product flow for lunches	5 days
28		Define ingredients needed	1 day
29		Stock kitchen	2 days
30		Inspect kitchen	3 days
31		**Packaging solution**	**16 days**
32		Define sandwich package	2 days
33		Define vegetable/fruit package	3 days
34		Define drink package	4 days
35		Design packages	5 days
36		Design box for lunch	2 days

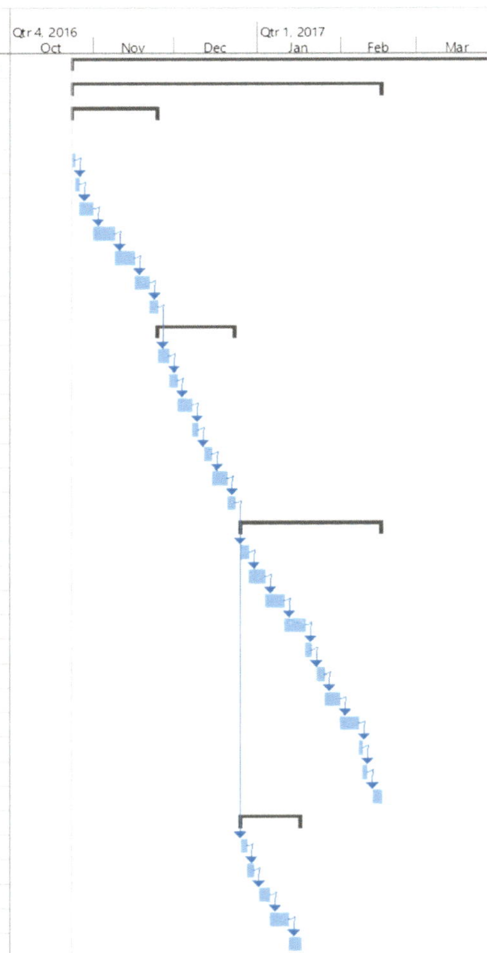

Figure 44: The Finley Project Schedule

These examples are provided in order to demonstrate the methodology of moving from a conception through the development of a specification and detailed schedule. These skills are critical for the success of any project manager, and a wide variety of approaches may be appropriate in order to generate predictive artifacts such as those provided. The next step for a project manager is to fully document these artifacts in a project charter.

The Project Charter

Prepared By:	
Version:	
Date:	

Sponsor:	
Project manager:	

Approval Signatures

Name____

Title______________________________

Date__________________

Name____

Title______________________________

Date__________________

Final Customer Sign Off*

Name____

Title______________________________

Date______________________________

Name____

Title______________________________

Date__________________

* Should be completed as a final signoff to the project.

Key Stakeholders			
Project Sponsor:		Phone:	
Division:		Department:	
Project manager:		Phone:	
Division:		Department:	
Colleges / Departments Affected:			

Summary:	How does this project contribute to the overall Vision of the Organization? With which strategy does this project align?

Problem/ Opportunity	What is the problem or opportunity that this project poses to correct or take advantage? (Not the project benefits)

Benefits	Identify the Tangible and Intangible benefits that will be derived from this project implementation.
Tangible: Intangible:	

Critical Success Factors	Critical success factors are those business factors that should be accomplished in order to consider this project a success.
1.	
2.	
3.	
4.	

In Scope	Describe the items that are in scope for the successful completion of this project. Any items not listed are not understood as being in scope.

Out of Scope	Describe the items that are out of scope for the successful completion of this project. Any items not listed are not understood as being out of scope.

Risks / Opportunities	What risks or opportunities are associated with completing this project, or not completing this project? What risks or opportunities can be identified when considering resources needed to complete this project?

Costs	What are the costs associated with this project (Cost are usually estimated with a confidence rating at the beginning and continue to firm up during the planning process.)

Costs:				
Hardware	$ -		Consulting	$
Software	$ -		Internal Labor	$
Support Contract	$ -			
Other	$ -			
Expense Total	$ -		**Labor Total**	$
	Total Cost:	$ -		
Initial Cost	$ -			
Monthly Cost of Ownership	$ -			
Monthly Benefit	$ -			
Cummulative ROI	#DIV/0!			
**Based on a three year rate of return*				

ROI is calculated as 1-[{Total Cost + (Cost of Ownership * 36)} / (Monthly Benefit * 36)]

Interdependencies	In order to complete this project, does it require any other projects or tasks outside of the Project Team's control to complete in order to be successful?

Milestones		In this section the user can identify the top 4 milestones and their proposed completion dates.
Milestone	Description	Completion Date
1.		
2.		
3.		
4.		

Technology	What technology will be used to deliver the solution?

Team Composition	How will the Project Team be composed?

Stakeholders	List all of the individuals, departments or users who will be impacted by the outcome of the project.

Resources	What resources are needed to complete this project in scope and on time? Resources include people, skills and equipment. Include an estimate to the time a resource is needed.

Assumptions	What assumptions are being made about the resources, budget, schedule, etc.?

Constraints	List major constraints that are on the project team and the completion of this project. Be sure to include milestone dates, resource availability and budget.

An Example Project Charter – Dinosaur World

Project Charter

Prepared By:	R. Monson
Version:	A
Date:	05MAR2017

Sponsor:	M. Crichton
Project manager:	R. Monson

Approval Signatures

Name____
Title_____________________________
Date_________________

Name____
Title_____________________________
Date_________________

Final Customer Sign Off*

Name____
Title_____________________________
Date_________________
Name____
Title_____________________________
Date_________________

* Should be completed as a final signoff to the project.

Key Stakeholders			
Project Sponsor:	R. FInley	Phone:	651-555-1234
Division:	Dinosaur	Department:	Dinoworld
Project manager:	R. Monson	Phone:	651-555-3241
Division:	Dinosaur	Department:	Design
Colleges / Departments Affected:	N/A		

Summary:	How does this project contribute to the overall Vision of the Organization? With which strategy does this project align?
Dinosaur World will be the first publically accessible theme park that will feature dinosaurs on exhibit. This park will be designed to resemble a theme park such as Disneyworld, but will feature the flora and fauna of the Jurassic period. No raptors will be included in the park.	

Problem/ Opportunity	What is the problem or opportunity that this project poses to correct or take advantage? (Not the project benefits)
Product marketing has determined that when the park is completed we will expect to host over 2 million visitors per year. With the current planned ticket prices and expected spending on food and other miscellaneous items within the park, this will translate into	

approximately $4 billion in annual revenue for the operation of the park. Additional opportunities will arise with the commercialization of products outside of the park. Expected costs of operation will not exceed $3 billion per year. In addition to this, researchers will be granted academic access to the park for research studies and subsequent publications, further publicizing the park.

Benefits	Identify the Tangible and Intangible benefits that will be derived from this project implementation.
Tangible: New revenue source for the organization Ability to expand into multiple parks offerings Unique offering that offered by other theme parks Transition of current dinosaur research to the field Intangible: Enhanced ability to develop and generate dinosaur offspring Increased understanding of care and feeding of dinosaurs Credibility and research associated with our current dinosaur development	

Critical Success Factors	Critical success factors are those business factors that should be accomplished in order to consider this project a success.
1.	Development of multiple species of dinosaurs
2.	Development of appropriate environmental factors such as flora
3.	Ability to care and feed for residents of the park
4.	Ability to operate theme park efficiently and safely

In Scope	Describe the items that are in scope for the successful completion of this project. Any items not listed are not understood as being in scope.

This project will include the acquisition of land, the development of appropriate buildings and enclosures as well as the design of tours and activities to be conducted within the park. This will require an understanding of the Park operations, and will address those activities that need to be conducted over appropriate satisfaction of the customers within park.

Out of Scope	Describe the items that are out of scope for the successful completion of this project. Any items not listed are not understood as being out of scope.

This project will not define the requirements for maintenance within the park. This project will also not address the issues locating and hiring appropriate personnel for the operation of the park, including the vetting of personnel for insane tendencies that might result in the generation of raptors or other carnivorous species, or their involvement with foreign governments or evil geniuses that may choose to purchase these creatures to the detriment of the United States.

Risks / Opportunities	What risks or opportunities are associated with completing this project, or not completing this project? What risks or opportunities can be identified when considering resources needed to complete this project?

There is a real concern that all species can coexist effectively

The development of a sustainable ecosystem is of concern
The current schedule shows to element must be completed within a period of 24 months
The current transportation infrastructure at the desired location for the park has not been developed as promised
The operation of the park will require additional lodging to be developed at the park site support those individuals not working first shift
Unspecified diseases or infections that may affect the flora and fauna of the park
The park introducing new strains of bacteria or infection into the general population

Costs	What are the costs associated with this project (Cost are usually estimated with a confidence rating at the beginning and continue to firm up during the planning process.)

				Data	
Task	Task 1	Task 2	Task 3	Cost	A
Dinosaur World	Dinosaur World Experience project	Plan Dinosaur world	Layout	6400	
			Design of features	9600	
			Layout of new park	16000	
			Layout of buildings for park	12800	
			Design of amphitheater	22400	
		Plan Dinosaur world Total		67200	
		Develop Buildings	Layout Bldg.1	9600	
			Layout Bldg. 2	16000	
			Layout Bldg. 3	19200	

				Detail plans for Bldg. 1	25600
				Detail plans for Bldg. 2	28800
				Detail Plans for Bldg. 3	9600
				Define walkways and traffic flow	12800
			Develop Buildings Total		121600
			Establish planned flow	Construct sidewalks	38400
				Construct railings	9600
				Construct Signage	16000
			Establish planned flow Total		64000
			Complete Dioramas	Purchase fossils	16000
				Complete statues of dinosaurs	22400
				Contract design of robotic dinosaurs	28800
				Install in park	9600
			Complete Dioramas Total		76800
			Design Park Tour	Layout Tour	19968
				Define timing of tour	4992
				Train guides	23296
			Design Park Tour Total		48256
			Implement Tours	Allow reservation planning	4992
				Collect tour in lobby	1664
				Execute tour through park	416
				Provide safety briefing	208

			Return through gift shop	208
			Provide gifts to tour personnel	208
			Conduct satisfaction survey	208
		Implement Tours Total		7904
	Dinosaur World Experience project Total			385760
Dinosaur World Total				385760
Grand Total				385760

Interdependencies	In order to complete this project, does it require any other projects or tasks outside of the Project Team's control to complete in order to be successful?

The current Jurassic research projects must be completed prior to implementation in the park. These include:

Development of different dinosaur species

- Diplodocus
- Stegosaurus
- Ankylosaurus
- Brachiosaurus
- Parasaurolophus

Development of flying dinosaurs
    Pteranodon
    Pterodactyl
Development of fish and sea subspecies
    Plesiosaur
    Elasmosaurus
Development of plant life for display
    Conifers
    Ginkgophytes
    Club mosses
    Ferns
    Sphenopsids
    Cycadophytes

Milestones	In this section the user can identify the top 4 milestones and their proposed completion dates.	
Milestone	Description	Completion Date
1.	Plan Park	April 14
2.	Develop Buildings	May 30
3.	Establish Planned Flow	June 24
4.	Complete Dioramas	July 29
5.	Design Park Tours	September 12
6.	Implement Tours	October 5

Technology	What technology will be used to deliver the solution?
Our proprietary technology for the development of plant life and animal life will be utilized extensively throughout the project. The	

development of the physical enclosures will be dependent upon information provide International zoos. Part design and development will be dependent upon consulting services provided by Disney Corporation.

Team Composition	How will the Project Team be composed?
The project will be led by R. Monson, With the following development team members: Thomas Amanda Bill Susan Sarah Steven James Other personnel will be utilized as necessary throughout the development phase.	

Stakeholders	List all of the individuals, departments or users who will be impacted by the outcome of the project.
Patrons	

Scientists Staff Vendors Landowners Government Insurance Flora and fauna Transport Investors Project team Energy development

Resources	What resources are needed to complete this project in scope and on time? Resources include people, skills and equipment. Include an estimate to the time a resource is needed.
Project team Dinosaur research results Plant research results Finley construction company (For park construction) Disney Corporation consulting Transportation authority Park locale governmental representation	

Assumptions	What assumptions are being made about the resources, budget, schedule, etc.?
The expense of producing the park will not exceed $4 billion	

including scientific research
The local community will accept the installation of the park
All necessary permits and approvals can be obtained
Director Spielberg can be prevented from entering the park premises at off hours during development for his filming of our work
The safety of the animals and park visitors can be maintained
The safety of operating personnel can be maintained
Expected patronage of the park will be realized

Constraints	List major constraints that are on the project team and the completion of this project. Be sure to include milestone dates, resource availability and budget.

The project will be completed by the scheduled park opening date
The animal population will be limited by the rate of development of new species in the dinosaur research laboratory
The plant population will be limited by the rate of development of plant species in the plant research laboratory growth
Safety of all personnel and animals will be maintained at all times
No new bacterial or infectious growths will be introduced to the general population

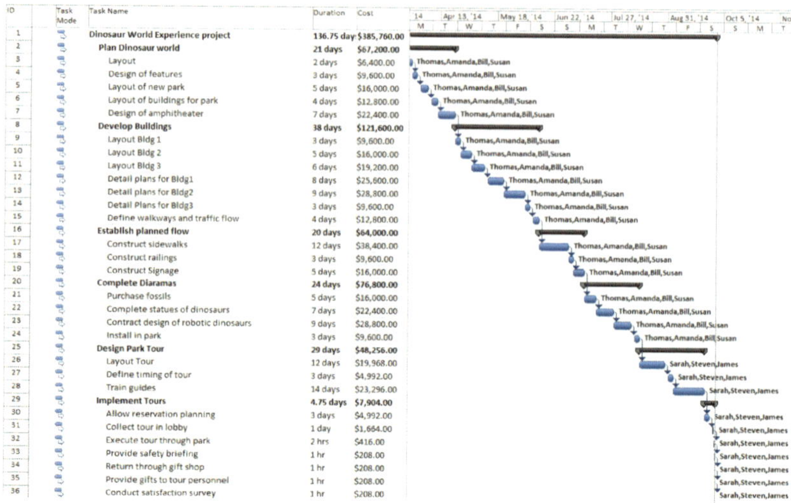

ID	Task Mode	Task Name	Duration	Cost
1		**Dinosaur World Experience project**	**136.75 day**	**$385,760.00**
2		**Plan Dinosaur world**	**21 days**	**$67,200.00**
3		Layout	2 days	$6,400.00
4		Design of features	3 days	$9,600.00
5		Layout of new park	5 days	$16,000.00
6		Layout of buildings for park	4 days	$12,800.00
7		Design of amphitheater	7 days	$22,400.00
8		**Develop Buildings**	**38 days**	**$121,600.00**
9		Layout Bldg 1	3 days	$9,600.00
10		Layout Bldg 2	5 days	$16,000.00
11		Layout Bldg 3	6 days	$19,200.00
12		Detail plans for Bldg1	8 days	$25,600.00
13		Detail plans for Bldg2	9 days	$28,800.00
14		Detail Plans for Bldg3	3 days	$9,600.00
15		Define walkways and traffic flow	4 days	$12,800.00
16		**Establish planned flow**	**20 days**	**$64,000.00**
17		Construct sidewalks	12 days	$38,400.00
18		Construct railings	3 days	$9,600.00
19		Construct Signage	5 days	$16,000.00
20		**Complete Diaramas**	**24 days**	**$76,800.00**
21		Purchase fossils	5 days	$16,000.00
22		Complete statues of dinosaurs	7 days	$22,400.00
23		Contract design of robotic dinosaurs	9 days	$28,800.00
24		Install in park	3 days	$9,600.00
25		**Design Park Tour**	**29 days**	**$48,256.00**
26		Layout Tour	12 days	$19,968.00
27		Define timing of tour	3 days	$4,992.00
28		Train guides	14 days	$23,296.00
29		**Implement Tours**	**4.75 days**	**$7,904.00**
30		Allow reservation planning	3 days	$4,992.00
31		Collect tour in lobby	1 day	$1,664.00
32		Execute tour through park	2 hrs	$416.00
33		Provide safety briefing	1 hr	$208.00
34		Return through gift shop	1 hr	$208.00
35		Provide gifts to tour personnel	1 hr	$208.00
36		Conduct satisfaction survey	1 hr	$208.00

Figure 45: Dinosaur World Schedule and Budget

Note that the detail project schedule and other supporting documentation should be attached to the project charter when completed.

To fully understand the Project Management life cycle, there are additional phases of the project that must be understood. These include Project Control, Project Execution and Project Close-out.

Phase 3: Project Control

Earned Value Management

In order to realistically manage a project or projects in an enterprise, it is beneficial to understand a practical and relevant manner in which to status a project. Many systems exist within organizations in order to accomplish status definition. In general, Earned Value Analysis provides significant predictive capability for the Project manager or Engineering leader, and is therefore a preferred method.

Earned Value Management Systems (EVMS) is a cost management technique wherein the value of completed work is compared to actual costs and the scheduled costs for the project. By examination of these two cost ratios, an excellent prediction may be created to define the expected outcome of the project.

Earned Value is typically expressed in terms of dollars, thereby allowing a comparison between completed work, timecards and actual costs on the project.

Earned Value is calculated by multiplying an estimate of the percentage complete for the task by the total estimated cost for that task. As an example:

Five weeklong test costing ten thousand dollars
Task is reported as thirty-seven percent complete
$0.37 \times 10,000 = \$3700$ of Earned Value

Therefore, in this instance the Project manager is allowed to take a $3700 credit for work that has been completed. This value would then be compared to the actual cost to complete the work, thereby giving some insight into cost. Secondly, this value will be compared to the work scheduled to be completed, thereby giving some insight into variance in the schedule.

So why would Earned Value Analysis be of any interest to the Project manager? It provides a significant ability to predict outcome of the project prior to completion. As an example, consider a $100,000 project that is currently underway:

We have spent $90,000
The earned value of the work completed is $80,000

This predicts that we will have a significant cost overruns on our project. Earned value allows us to answer the next question to be asked regarding this project; what is the amount of the cost overrun? To establish this, we must think through what this information provides us regarding our current status.

We have $20,000 of work remaining
We have $10,000 of funding remaining
We know we are accomplishing our work at and 8/9 efficiency
Therefore we will require approximately $22,000 to complete the work
Therefore the expected cost overrun is $12,000

And identical analysis can be performed related to the schedule status as well.

Several variables are necessary to establish the earned value of the work. They include:

BCWP – budgeted cost of work performed
ACWP – actual cost of work performed
BCWS– Budgeted cost of work scheduled

In order to establish metrics for our project, we are interested principally in two factors. These include the cost variance, relating to the efficiency with which we are completing our work, and the schedule variance, relating to the rate at which we are completing our work. With these two pieces of information we are enabled to make an accurate prediction regarding the final outcome of the project once the project is underway.

CVI (Cost Variance Index) = BCWP/ACWP
SVI (Schedule Variance Index) = BCWP/BCWS

Both of these variables are utilized in a similar manner when assessing the status of the project. If we utilize the cost variance index as an example, it is obvious that if the budgeted cost of work performed is equivalent to the actual cost of work performed; we have a CVI equivalent to 1.0. This implies that we are paying as much for the work as we predicted we would pay for the work accomplished. This implies that we will finish the project on budget.

When we consider how projects are normally evaluated, we see the Gantt chart utilized as the status tool. This is shown in Figure 46.

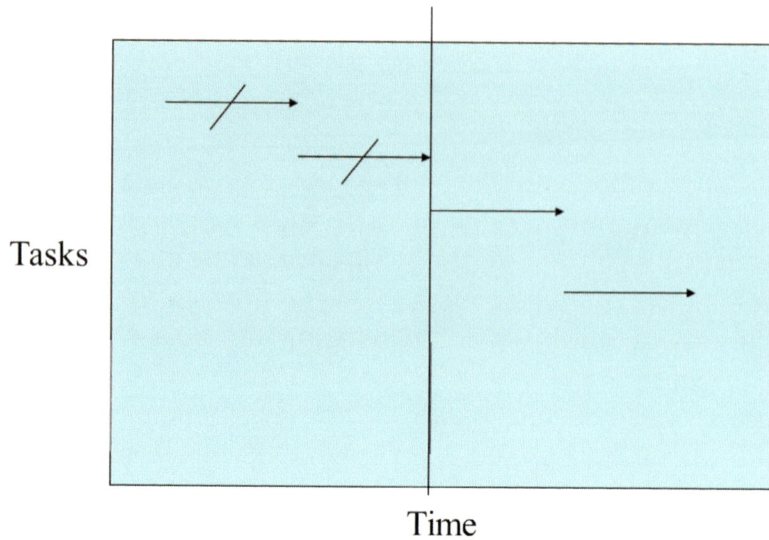

Figure 46: Gantt Chart Status

As can be seen, our tasks are listed on the project plan, two of which have been completed and with a vertical line representing the current date. Most project managers would assess this project as being on schedule, but they may not have considered several interesting factors related to the status of the project:

The individual assigned the task supplied the status.
It may be of interest to the associate to report work is complete that they fully expect will be completed in the near future.
There is considerable question as to what the definition of "done" means.

By virtue of these considerations, it is potentially possible that although the project appears to be on schedule utilizing the Gantt chart, it might be off schedule.

Additionally, if the third task were to be reported as thirty-seven percent complete, we could consider the earned value of that specific task in our analysis. But in the big scheme of things, this means one thing to the project manager: the task is not complete. Therefore, we see that if the tasks are not sufficiently aggregated, any reporting approach may provide us with less than desirable data. In general, it is advised that no task should exceed the amount of error that you can support in your project tracking. Therefore, many project managers will set a rule that no task may exceed forty hours in duration, or something similar.

The second most commonly used tracking tool for project managers is the current budget. We can make a similar inaccuracy (or non-representative) argument relating to the consumption of the budget during the execution of the project. Remember, without a complete understanding of how the consumption of the budget relates to the completion of the work remaining, it is a dubious metric to be considered.

Rather than use these typical metrics for our project assessment, we instead will use two EVMS metrics in order to make a project prediction.

Is the work getting completed at close to the estimated cost?
Is the work progressing as originally scheduled?

From this we will predict if we can meet our specification, schedule, and budget.

Cost variance is a relatively simple concept, and it describes the deviation of actual costs from expected (or budgeted) costs. This can be seen in Figure 47.

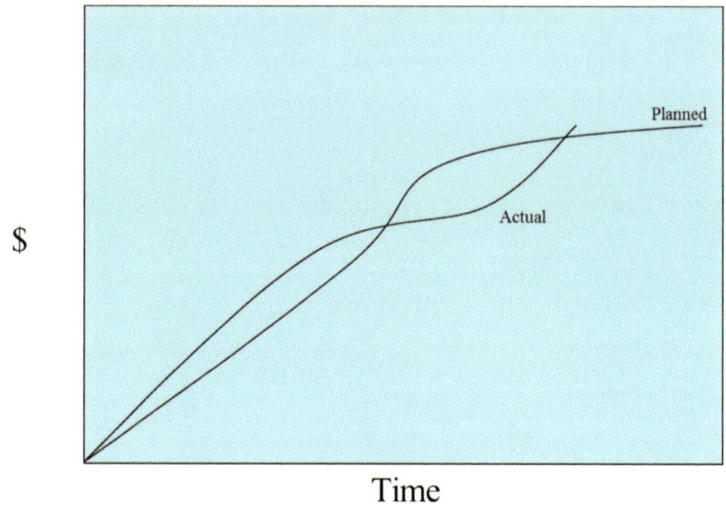

Figure 47: Cost Variance

In this example, the actual costs are shown to have outpaced the expected costs throughout the first half of the project, and now are significantly less than the expected costs. The real challenge lies in deciding how we would utilize this information to make a prediction of the actual outcome of the project. One can see that many extrapolations of the information could plausibly be used, so which is the most predictive?

Schedule variance is related to the rate at which work is being completed on the project. It is in essence a comparison between the tasks that are completed at this point in time, and what tasks were expected to have been completed at this point in time. The comparison of these two values enables us to gain some insight into the expected outcome of the schedule for the project. This is shown in Figure 48.

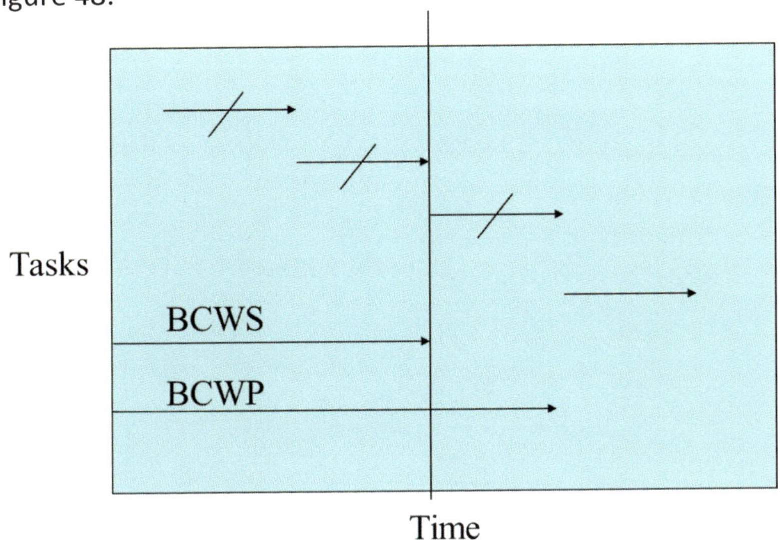

Figure 48: Schedule Variance

These variances can be of significant importance to us, but we must also understand the dynamic nature of the project environment. Regardless of the appropriateness or accuracy of the metrics used in the tracking of a project, due to a wide variety of considerations within the organization the actual path of the project is very likely to deviate significantly from the expected path. Therefore, it is likely that we will continuously deviate from the planned path during the execution of the project. We must consider these variances with regard to the appropriate control limits as the project progresses. Thus, we will consider that if the project is within a given percentage of the expected position, we will limit our interaction with the project team in order to minimize the potential Chaos interjected into the project environment. The multiple paths that different development teams may follow in the completion of the project and appropriate control limits are shown in Figure 49.

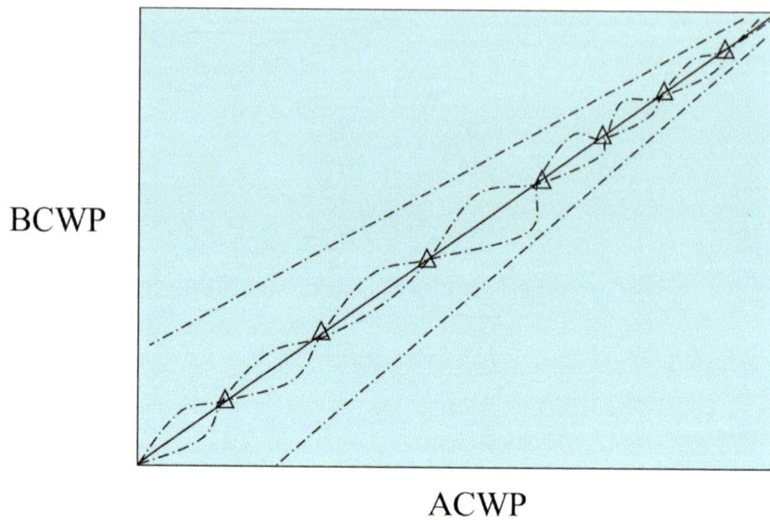

Figure 49: Control Limits for Cost Variance

Therefore, we will consider both the cost and schedule variances in our typical assessment of project status. As an example, in the following diagram we are reporting a cost variance of 1.1 and a schedule variance of 1.0. This is shown in Figure 50.

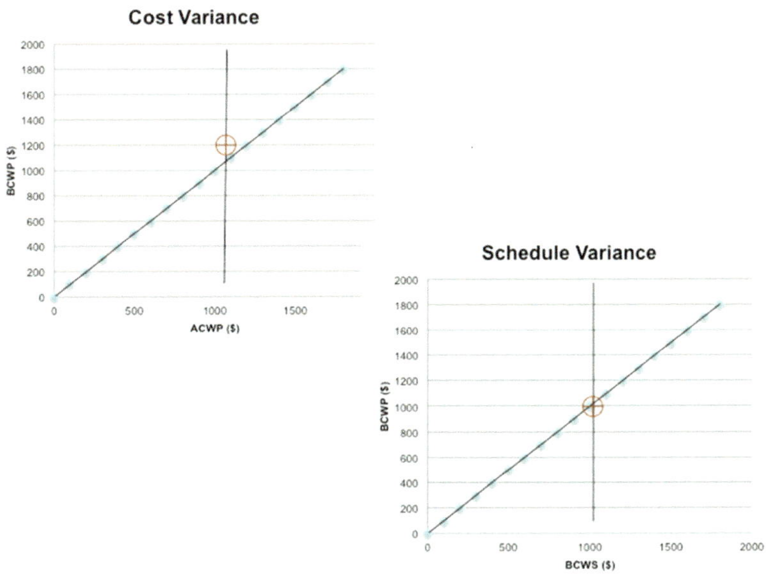

Figure 50: Project Status Charts

These charts show that the work is progressing as expected, and therefore it is likely we will meet our original schedule. The cost variance has a favorable value, and we would predict that we would deliver this project under budget by approximately ten percent. These results are extremely straightforward.

In a more sophisticated example, we have reported a cost variance of 1.0 and a schedule variance of 0.5. This is shown in Figure 51.

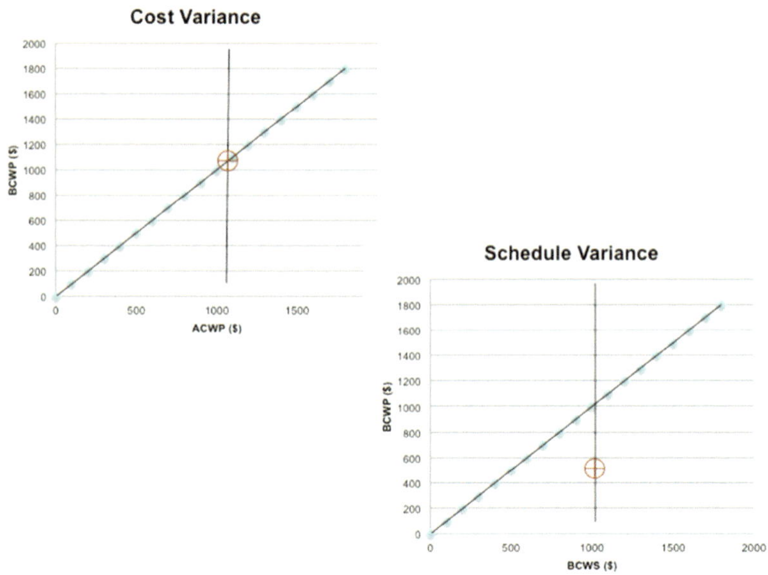

Figure 51: Project Status 2

This project has a much more interesting story to tell. It appears from our cost variance that we will be able to complete our project within the specified budget. However, we have a significant problem relating to our schedule variance; only half of the expected work is completed at this time. In order to get this project back on schedule, we need to add people to our team in order to accomplish more work at a quicker rate. It is possible to examine the schedule variance and make a prediction of how many personnel are needed. In this case, we have spent one-half of the time and completed one-quarter of the work. In the remaining one-half of the time, we had three-quarters of the work yet to complete; therefore, we must triple our current workforce in order to meet our original deadline.

One concern that is frequently voiced by project managers is that the addition of significant personnel to the project team will cause significant cost overruns. While that may be true due to inefficiencies that are inherent in the expansion of any work team, in this particular case it is not a significant concern. In fact, the adjustment of the workforce in order to adjust the schedule variance will typically have little or no bearing on the cost variance.

As an example, let us assume that we have hired a neighborhood boy to mow our lawn. Normally this requires three hours of effort and he is rewarded with $30 payment. This results in a $10 per hour rate for completion of the job. Imagine that we have scheduled a significant event on our lawn one day, and the individual responsible for mowing the lawn no longer has three hours in order to complete the work. If we request he complete the work in one hour instead of the usual three, (we would like the work completed prior to the event,) then he must increase his workforce. In this case, let us assume that he brings along two friends to assist in the mowing of the lawn. As might be expected, the work is completed in one hour. The payment is still $30, which in this case will be paid to three individuals, and still results in $10 per hour as the rate of pay. Therefore, the cost variance has not changed even though the workforce has been tripled in size. This of course ignores the potential problems associated with multiple personnel mowing the lawn simultaneously.

In most organizations, the EVMS information is not expressed in a manner as simple as the previous two diagrams. In fact, it will look more like the data presented in Figure 52.

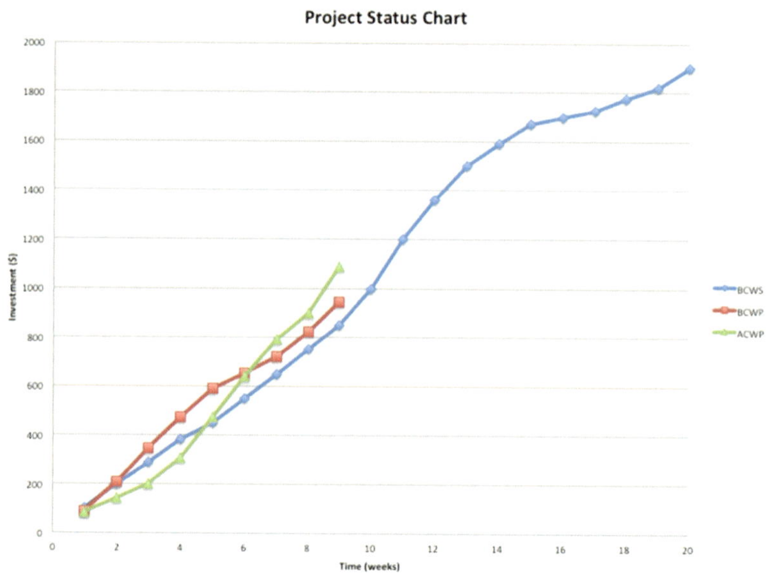

Figure 52: Project Status 3

While the same information is present on the chart, this method of reporting can be quite problematic with regard to predicting the outcome of the project and managing it accordingly. It may be beneficial to reconstruct diagrams from this format to those described in the previous diagrams. This will alleviate the chance of misinterpreting the information on the chart, as it requires several translations of the data in the reader's head. If this is not possible, it is often better to simply report the ratio values rather than a graphical chart.

Regardless of the method selected, the user must utilize this information to decide if action must be taken to refocus efforts on the project. This can be defined by meticulous analysis of the EVMS ratios, but fortunately there is a much better way. This simplified method is to utilize the system of a bull's-eye chart.

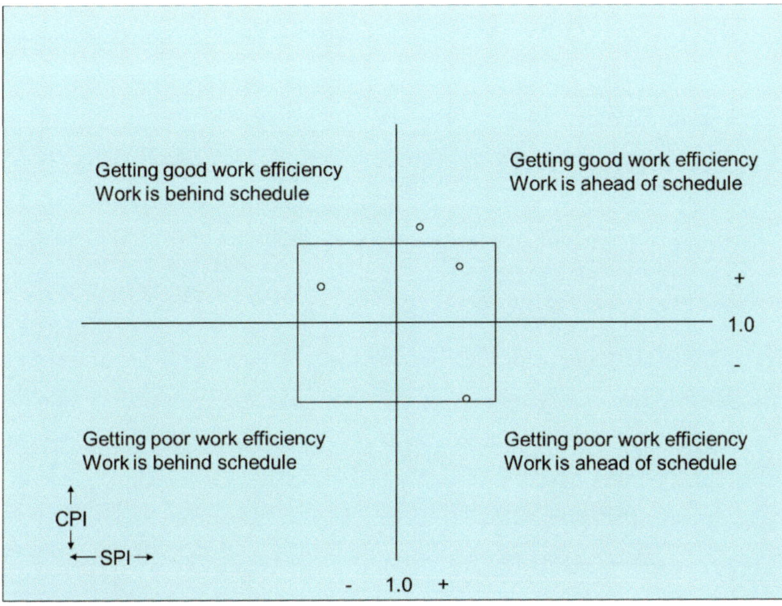

Figure 53: Bulls-eye Chart

In a bull's-eye chart, the cost variance index and schedule variance index are located on the Y-axis and X-axis respectively. As can be seen in Figure 53, this places the ideal performance of the project residing at the bull's-eye in the center of the chart. A control box may also be supplied on this chart as a reference to establish where the desired performance for the project resides. This control box is equivalent to the control limits previously discussed.

Data can then be provided on this bull's-eye chart over a period of time, creating a longitudinal view of the status of the project. This gives the project manager a quick insight as to the status of the project, with little danger of misinterpretation. Additionally, criteria for assessment can be included on the bull's-eye chart, providing general recommendations to the Project manager. This can be seen in Figure 54.

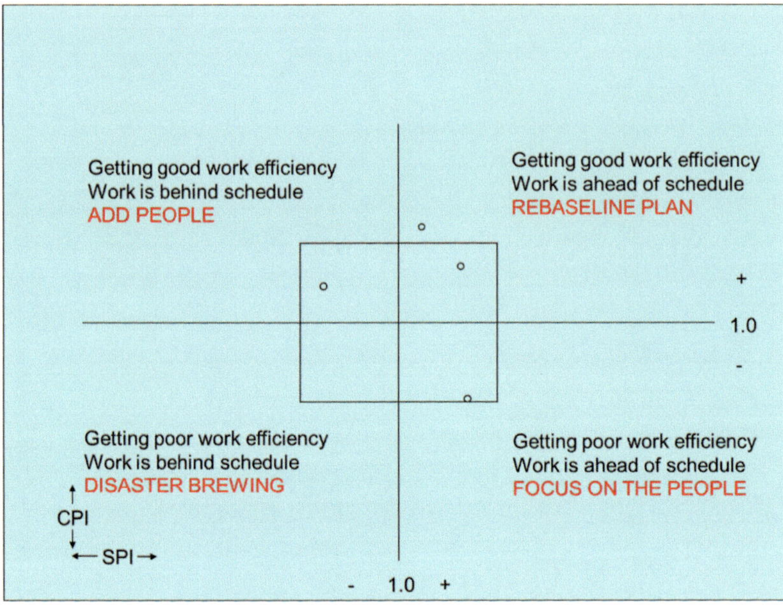

Figure 54: Bulls-eye Chart with Recommendations

An example of a real bull's-eye chart with twelve months of status data posted is shown in Figure 55. Note the color coding of the chart; showing green, yellow and red zones of performance.

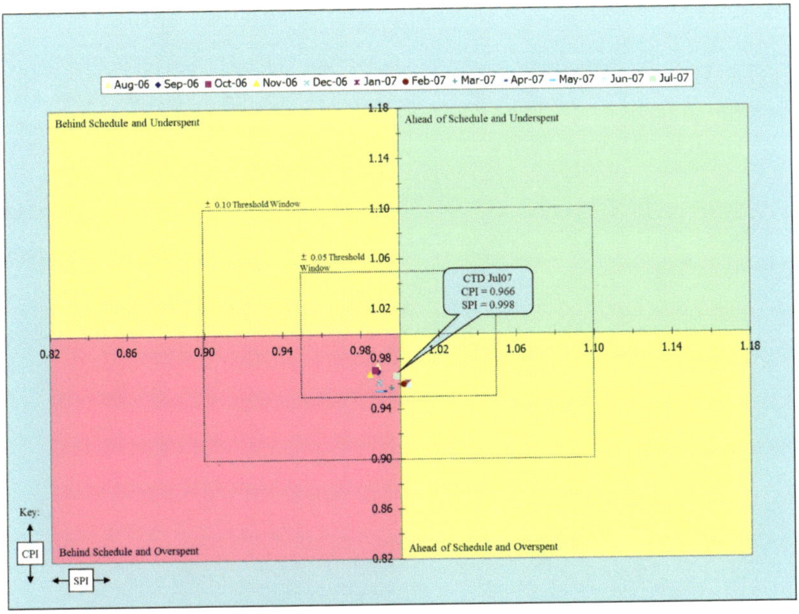

Figure 55: Example Bulls-eye Chart with Twelve Months of Data

Extrapolation of the data provided by the cost variance and schedule variance is often used to make a prediction of project outcome. The methods by which this extrapolation is accomplished are shown in Figure 56. Note that this is using the total project performance to define the predicted outcome, as past performance is indicative of future performance. This is normally the best prediction for a project manager to use when extrapolating. One caveat should be mentioned; early in the project the data can be very volatile, so exercise caution when predicting outcome based upon less than twenty percent project completion.

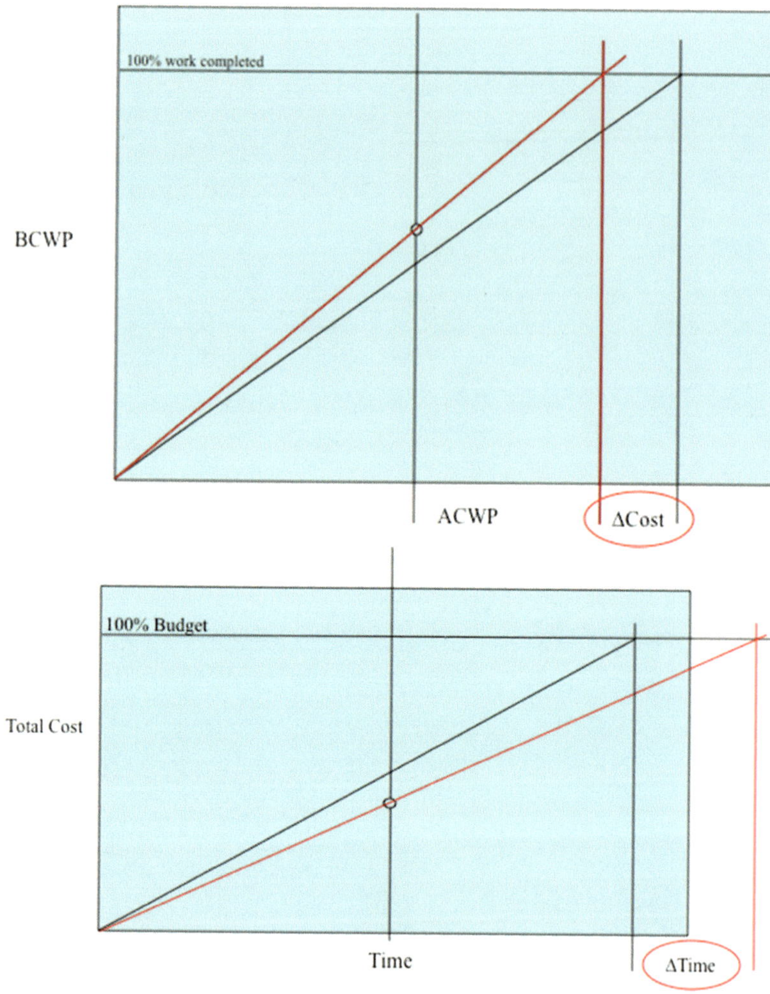

Figure 56: Extrapolation of Project Outcome

Since it is necessary to reach one hundred percent of the budgeted cost of work completed (BCWP) in order to finish the project, we utilize this line as our limit for prediction as shown in these two diagrams.

It is extremely important for any engineering leader to understand how to quickly assess the status of a project given these two variables and define appropriate actions forward to enable maximum probability of project success. Several examples are provided here for consideration.

Predicting Project Outcomes

Your project is progressing reasonably well, but you are worried about the reports you have received on the earned value of your tasks. You have gotten differing results over the past three weeks, and you are very concerned about what this means to the project. Assuming you are halfway into the project, what do you report for each of the following scenarios?

CVI = 1.1, SVI = 1.0
First off, CVI is greater than one, so we know we are getting more work done than we estimated per dollar; our efficiency is good. Therefore, we know we should finish ahead of budget. The SVI shows we are spending at the right rate, so we should finish on time, as well.

CVI = .85, SVI = 1.2
The CVI says we are not as efficient (our estimates were optimistic) as we planned, so we should now predict the amount of overrun based on a reduced efficiency, a fifteen percent overrun. SVI tells us we are ahead of schedule, which might be a good argument for us to ask for further funding, as it appears we will finish before the projected end date.

CVI = 1.1, SVI = .95
The cost variance is favorable, but the schedule variance is unfavorable. They are both reasonably close to 1.0, so we can probably make up the schedule variance by applying

additional resources. This may reduce efficiency, bringing CVI closer to 1.0.

A typical project status spreadsheet as seen in many organizations is shown in Figure 57. Note that all of the previously described information is present in this spreadsheet, in addition to several different assessment methods that are utilized to calculate related variables with regard to project status. These variables include Budget At Complete (BAC), Estimate To Complete (ETC), Estimate At Complete (EAC), Variance At Complete (VAC), and others (see Figure 58.) The interested reader can find much more information available at www.DAU.mil.

	Jan-03 $	Feb-03 $	Mar-03 $	Apr-03 $	May-03 $	Jun-03 $
BCWS	122,148	263,656	417,065	597,229	811,730	1,043,605
ACWP	138,901	296,993	459,920	654,854	888,258	1,132,596
BCWP	136,064	237,015	373,080	511,339	730,798	959,035
BAC	2,194,588	2,194,588	2,194,588	2,194,588	2,194,588	2,194,588
CV	-2,837	-59,978	-86,840	-143,515	-157,461	-173,561
CV %	-2.09%	-25.31%	-23.28%	-28.07%	-21.55%	-18.10%
CPI	0.980	0.798	0.811	0.781	0.823	0.847
SV	13,916	-26,640	-43,985	-85,890	-80,932	-84,570
SV %	11.39%	-10.10%	-10.55%	-14.38%	-9.97%	-8.10%
SPI	1.114	0.899	0.895	0.856	0.900	0.919
EAC 1	2,240,346	2,749,940	2,705,414	2,810,533	2,667,443	2,591,754
EAC2	2,025,419	3,025,646	2,970,149	3,172,625	2,864,478	2,720,426
VAC 1	-45,758	-555,352	-510,826	-615,945	-472,855	-397,166
VAC 1 %	-2.09%	-25.31%	-23.28%	-28.07%	-21.55%	-18.10%
VAC 2	169,169	-831,058	-775,561	-978,037	-669,890	-525,838
VAC 2 %	7.71%	-37.87%	-35.34%	-44.57%	-30.52%	-23.96%
ETC	1,886,517	2,728,653	2,510,228	2,517,771	1,976,220	1,587,830
ECD	19-Nov-03	5-Feb-04	7-Feb-04	21-Feb-04	5-Feb-04	29-Jan-04

Figure 57: Project Status Spreadsheet

Earned Value Terminology

EVMS	–	Earned Value Management System
WBS	–	Work Breakdown Structure
OBS	–	Organizational Breakdown Structure
Control Account	–	Intersection of WBS and OBS where costs are accumulated and performance measured
Work Package	–	Subdivision of work within a Cost Account Provides level for Earned Value assessment
IMP	–	Integrated Master Plan (Event, Accomplishment, Criteria)
IMS	–	Integrated Master Schedule (IMP + Task)
CP	–	Critical Path: A continuous path of activities in a network having the longest total duration and thus defining the minimum duration of the project. The path with zero float.
Forward Pass	–	Determines EARLY DATES for each activity – the earliest each activity can be done, based on logic and duration
Backward Pass	–	Determines LATE DATES for each activity – the latest each activity can be done without impacting the project end date, based on logic and duration
Total Float	–	The amount of time an activity can be delayed or expanded before it impacts the Project Finish Date
BCWS	–	Budgeted Cost of Work Scheduled
BCWP	–	Budgeted Cost of Work Performed (Earned Value)
ACWP	–	Actual Cost of Work Performed
BAC	–	Budget at Completion
EAC	–	Estimate at Completion (ACWP + ETC)
ETC/CTC	–	Estimate to Completion/Cost to Complete
LRE	–	Latest Revised Estimate (see also EAC)
VAC	–	Variance at Completion (BAC – EAC)

Sample IMP/IMS Structure

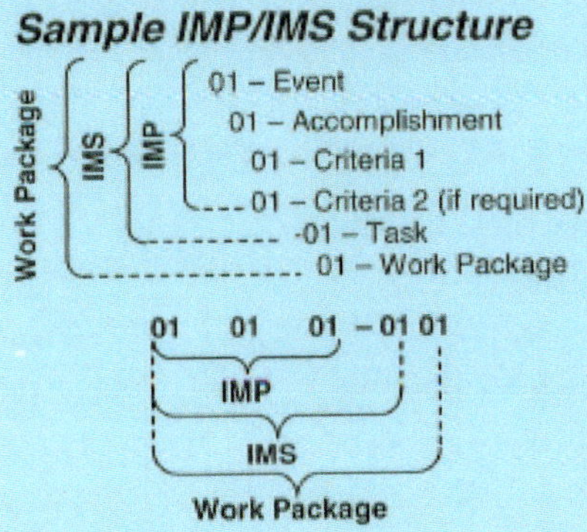

2005

Figure 58: EVMS Terminology

Earned Value Formulas

Cost Variance (CV)

CV = BCWP - ACWP
(60) = 1520 - 1580

$$CV\% = \frac{CV}{BCWP} = \frac{(60)}{1520} = (4\%)$$ 60 units overrun

Schedule Variance (SV)

SV = BCWP - BCWS
30 = 1520 - 1490

$$SV\% = \frac{SV}{BCWS} = \frac{30}{1490} = 2\%$$ 30 units ahead of schedule

Cost Performance Index (Efficiency)

$$CPI = \frac{BCWP}{ACWP}; \frac{1520}{1580} = 0.96$$

Performed 0.96 units for every 1.00 unit incurred.
Index > 1.0 = under cost
Index < 1.0 = over cost

Schedule Performance Index (Efficiency)

$$SPI = \frac{BCWP}{BCWS}; \frac{1520}{1490} = 1.02$$

Approximately 2% ahead of schedule
Index > 1.0 = Ahead of schedule
Index < 1.0 = Behind schedule

To-Complete Performance Index

TCPI is related to CPI and tolerance is within ±5%.

$$TCPI(LRE) = \frac{BAC - BCWP}{LRE - ACWP} = \frac{\text{Working Remaining}}{CTC}$$

TCPI measures the to - go performance efficiency required to meet the declared LRE.

$$TCPI(LRE) = \frac{5000 - 1520}{5100 - 1580} = \frac{3480}{3520} = 0.99$$

Estimate at Completion (EAC) Formulas

1 Weighted Cost/Schedule EAC (80/20) Method

$$EAC = Actuals + \left[\frac{BAC - BCWP_{cum}}{(.8 \times CPI) + (.2 \times SPI)}\right]$$

EAC projection based on past cost and schedule performance.

2 Cum CPI Method

$$EAC = \frac{BAC}{CPI_{cum}}$$

Most optimistic method. Assumes program will perform the same in the future as in the past.

Earned Value Management Data

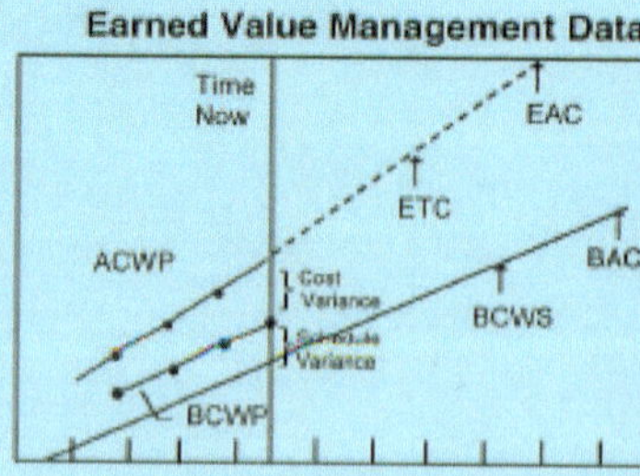

2005

Figure 59: EVMS Formulas

In the event you find a challenge that cannot be met in terms of internal project manipulation, you then resort to Trade-Off Analysis.

Trade-Off Analysis

Successful Project Management is an art and a science. In some cases the project manager finds himself in a situation where the time, cost and performance issues are impossible to maintain without project plan modifications. Here a Trade-Off analysis is used.

Six Steps to Project Reconciliation:

1. Recognize and understand the basis for project conflicts
   Human Errors, uncertain problems,
   Unexpected problems
2. Review the project objectives
   From the perspectives of all stakeholders
3. Analyze the project environment and status
   Detailed analysis of planned/actual cost/time/performance (EVA)
   Assessment of potential resolutions
4. Identify possible courses of action
   Brainstorming / Discussions
5. Select the best alternative
   Define best plan of action
   Define impact on Cost/Time/Performance
   Negotiate with Stakeholders for new plan constraints
6. Revise the project plan and publish it to all stakeholders
   Maintain signature approvals as required

The Project Simulation

The Project Simulation was created in order to produce an environment where project management students could run sample projects in order to get familiar with the use of project metrics and how the actions of the project manager can impact the end results of the project. It is designed to allow the rapid generation of randomized projects, including both schedules and personnel for assignment.

To begin, download The Project Simulation from the webpage at www.mninter.net/~rjmonson

The simulation is enclosed within an Excel spreadsheet, and contains macros for use during the simulation. Open the file and enable the use of macros in order to begin the simulation.

When the spreadsheet is first opened, it will contain the last executed and saved project, and will appear as seen in Figure 60.

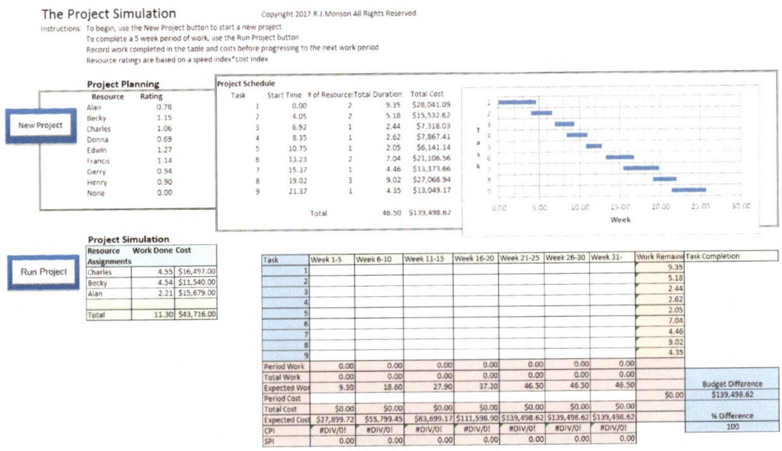

Figure 60: Opening Screen of Simulation

This opening screen shows the eight resources available for assignment on the project, and a rating associated with each person. This rating is a value calculated from the product of the individual's performance index and his or her cost index. The project manager will not know exactly how the individual performs relative to cost until he or she has had a chance to observe the outcomes of assignments. It is assumed that all personnel can accomplish all tasks equally well.

Next to the resources, a project schedule has been defined that has nine tasks; with the start dates for each task relative to an arbitrary start for the project, measured in weeks. The number of resources required for each task is listed, and the cumulative work weeks needed are also listed along with a fixed cost for the task, calculated from an average billing rate for personnel. The Gantt chart for the project is also shown for the review of the project manager.

This project and resource plan is created by utilizing the Run Project button to the left of the planning boxes. By pressing this button, the project manager resets all project variables, creating a new resource list, schedule and budget. Once created, this plan will remain static for the duration of the project.

In the lower section of the spreadsheet, we have a Run Project button. This is associated with the execution phase of the project. In order to utilize this function and begin a project, the project manager must decide what resources will be assigned for the work period. You may assign as many as three resources at any time, and they will be assigned for the entire duration of the time cycle, five weeks. The assignment of three individuals is shown in Figure 61.

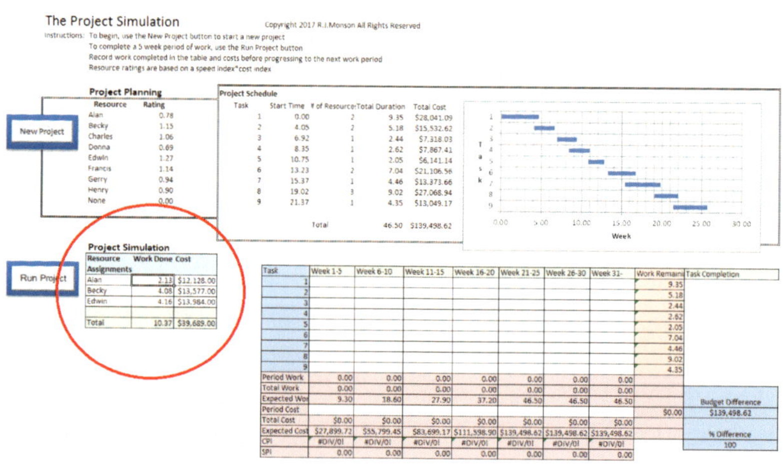

Figure 61: Resources Assigned

The project manager then pushes the Run Project button to see how much work was accomplished during the time period. In this case, 10.37 weeks of work was done, and so the project manager can choose to apply it to any tasks, making sure the amount applied does not exceed the 10.37 accomplished. In addition, the project manager must insert the cost of the work completed into the appropriate cell in the project execution worksheet. These records are shown in Figure 62.

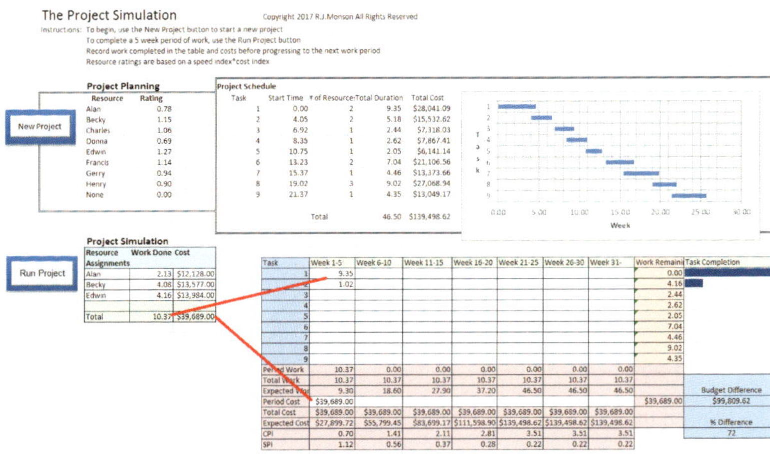

Figure 62: Transfer of Work Completed to Worksheet

The project manager can now see how much of each task is completed, how much budget remains, as well as the Earned Value metrics for the project, the Cost Performance Index (CPI) and the Schedule Performance Index (SPI). In addition, the project manager can assess what work each resource provided for his or her respective cost. Note that the work accomplished each time period will change for resources, but their costs and ratings will not. Thus, the project manager must decide what is important in the selection of resources.

As the project progresses, the project manager will continue to record results in the project worksheet until all the work is completed. At completion, the project manager can note the total duration of the project and the actual costs as compared to budgeted costs. This is shown in Figure 63.

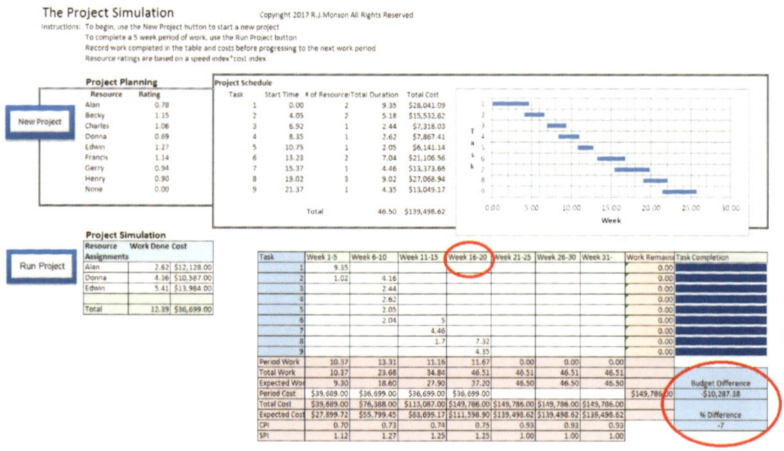

Figure 63: Completed Project

In this case, the project manager was able to complete the project in just twenty weeks (five + weeks ahead of schedule), and at a cost of $149,768.00, which was seven percent over budget.

The student may repeat this simulation as often as needed in order to enhance familiarity with the methods and metrics used in the execution phase of the project.

Phase 4: Project Execution

Project Execution is probably the most convoluted and challenging phase of any project endeavor. This is the portion of the project in which the project leader must rely upon his or her ability to provide an engaging project plan that will result in the employees investing their efforts in the successful completion of the plan. This covers a wide variety of topics, including individual motivation, communications, social and personality styles, group behavior, sociology, psychology and anthropology. In general, technical professionals gain these skills through application of the project techniques previously mentioned, and often must learn by observation in the real world. It is of great benefit to the project leader, however, to investigate the wide variety of publications available to assist in this area of project execution.

The Truths of Management

When undertaking the management of a team, it is crucial for the project manager to understand the role that people play in the success of the project. Fundamentally, this means that the project team members largely control the outcome of the project. More importantly, if they choose to make sure that the project fails, it is very simple for them to cause that failure while appearing to work effectively. On the flip side, if they choose to make the project succeed, it is imperative that all members of the project team are focused on the success of the project.

For this reason, it is important for the project manager to pay close attention to both the individual needs and group dynamics of the project team throughout the duration of the development. There are a few simple rules that will enable the project manager to better understand his or her project team, and these are referred to as the "Truths of Management."

When we consider the five-phase project life cycle, beginning with project initiation and project planning, we discover that the lion's share of the time and resources consumed by the project will occur during the project execution phase. In general, this phase will consume between eighty and ninety-five percent of the project budget. For that reason, it makes perfect sense that the project manager should become highly adept at managing this particular phase, as the other phases will have a smaller impact on the project overall.

When beginning a new project or when bringing a new team together for the first time, we often spend some time dealing with introductions and getting to know one another. This is clearly not a superfluous activity, as it is crucial to the initial formation of the team that the project manager spends some time getting to know the individuals. Why would this be important? Clearly, the enhancement of communications and the development of relationships on the project team will result in a higher success rate for the project overall. Therefore, it is useful for the project manager to discuss various topics with the team members, such that he or she may understand the communication strengths and weaknesses of the team. The inter-team interactions and relationships will need to be created through paying attention to the agendas and attitudes (and the various perspectives) the project team members bring to the table.

So let us imagine that we as project managers must answer a relatively simple question:

How much productivity can I expect from a person in a single day?

First, we should consider why this question is important. As project managers, we will plan large efforts that require huge amounts of resources within the organization. If you create a project plan that will consume something on the order of ten thousand person hours within the organization, it is crucial that you understand how much productivity can be expected from an individual within the organization, as this will define the duration of that project.

So how precisely do we need to know this value?

Perhaps it would be more appropriate to ask how precisely we need to predict the outcome of the project.

This must be known with a high degree of accuracy, probably within ten percent.

We also are aware that different people will provide different levels of productivity, and they will provide different levels of productivity on different days. By virtue of this, we will not be able to project with any degree of certainty the overall level of productivity on our project. While this is not good news, it is important that we recognize that this is the nature of the work we undertake when we manage projects.

This establishes for us our first truth:

1) When dealing with people, the correct response regarding their predicted behavior is usually: it depends.

This sets the stage for us to understand that we simply cannot know with any degree of certainty what the productivity will be with regard to our project team. This is terrible news for the project manager, whose responsibility is to predict accurately the outcome of the project. But, not all is lost.

If the project manager focuses carefully upon the job satisfaction of the project team members, it enables a much higher degree of predictability with regard to the performance of the project team. Higher job satisfaction will result in higher productivity, lower turnover and higher engagement on the project. This implies that if we focus on taking care of the people, in general, they will take care of the work.

This is a highly attractive theory from the perspective of the autonomy of the individual within the workplace, but it also implies that the project manager must be comfortable with allowing the project to be almost completely out of his or her control. This requires a great deal of effort for most project managers at the start of their project management careers. Nonetheless, the project manager must recognize that the team must be allowed to operate autonomously with appropriate guidance, as this will generate the highest productivity level.

This raises more questions:

How do we get the project team members to be on the

team?
How do we get predictability in the performance of our project team members?

To get the members on the team fully engaged in the effort, we must create a common understanding of the project itself. This implies we have to answer those three questions previously discussed:

What are you going to do?
How long will it take?
What will it cost?

These questions are typically answered by the generation of the detailed project plan, as it defines the constraints to be addressed by the development during the execution of the project. However, it does not always provide a compelling argument for the majority project team members. For this reason, we must spend considerable time thinking about how we can establish the importance and value of the project in terms that each of the project team members will appreciate. This implies that we understand the team members, and that we can align their individual needs and goals with those of the project.

Another challenging factor for most project managers is the inability to identify exactly when project begins. So let us take a simple example: imagine that your supervisor enters your office and asks you for a recommendation with regard to a problem that has been challenging the manufacturing floor for some period of time. Most individuals would perceive this as a supervisor asking for a simple opinion with regard to the problem. In reality, if we consider the needs of the supervisor, we will realize that he or she must provide a solution to a challenging problem in a manner beneficial to the organization.

This means that the supervisor must provide not only an appropriate approach to resolve the problem, but also a recommendation that includes the resources to be consumed in the project and an expected completion date. These factors are familiar to project managers.

Basically, this means that when a supervisor asks for a recommendation, he or she is asking for a project plan regarding the proposed solution to the given problem. A good project manager recognizes this immediately, and provides not just an opinion regarding the challenge identified, but also a perspective on the expected cost and duration of the project, based upon the project plan he or she has created.

If the project manager does not recognize this is the beginning of a project, he or she may provide an opinion to the supervisor suggesting what needs to be accomplished on the project without an appropriate budget or schedule. By virtue of this, the supervisor may bring forward the proposed solution to the problem and without other background information, may be required to generate a budget and end date. Unfortunately, these managerial-created budgets and schedules often do not reflect reality, largely because by the subject matter's expert, the project manager, did not create them.

This phenomenon does, however, clarify that if we are not proactive in our project planning, we are likely to get exactly what we deserve with regard to the future expectations of the projects we are assigned. Therefore, managers must recognize that projects begin very early, and the proactive project manager often sets the stage for success more effectively than the reactive project manager.

We can consider if we were taught how to properly manage projects in our educational background. To answer this question we don't really need to go any further than to ask:

Did you ever leave a term paper for the last night?

If you are like most people, this is something that has occurred in your past. We can ask ourselves:

Why would this be considered acceptable practice?

It turns out the answer is quite simple:

Because we can.

At some point in our careers we found ourselves in a situation when we did not have enough time to properly complete a term paper. In that particular case, we did the best we could, and we short-circuited the process we have been taught in order to be able to produce a practical product in the time remaining to us. As it turns out, most of us turned in that term paper and we ended up getting an acceptable grade for the work. This, then, supported our belief that in fact we do not need to follow all of the appropriate planning and execution rules taught to us in school regarding the development of a term paper.

Upon consideration, it is obvious that we were taught an appropriate way to develop a complex product like a term paper. We were taught to begin by outlining the paper and producing a rough draft, editing the draft to produce a second draft, and finally, rewriting the second draft completely to produce a final draft. This careful process is designed so that the final draft is significantly better than the rough draft. However, if you turned in a rough draft and received an acceptable grade during your educational career, you learned that the project management rules that were outlined for the development of the paper did not apply to you. This is the fundamental problem many project managers face.

We realize that there are appropriate ways to complete our work and that we should be practical in the manner in which we approach it. As our work becomes more complex and comprehensive, these methodologies and procedures become much more critical for our potential success. While in the past we could focus on minimizing our input in order to maximize our output, with more complex work we know that minimizing the input results in a lack of output and could also potentially result in considerably more work for the project team. Therefore, with more complexity in our work, the methodologies of project management become ever more critical for our future success.

We can ask ourselves another simple question that helps to illustrate some of the challenges we face in the project management domain:

What makes you successful in project management?

Many people would respond by saying that if we manage to deliver our specification schedule and budget we are successful. In reality, however, we know that we can deliver our specification, schedule and budget, and still fail in our project. This might be due to our leaving a trail of bodies behind us within the organization, or a lack of insight into the expectations of our sponsor within the project team. Likewise, we may succeed when we do not meet our original specification schedule and budget if we focus on accomplishing the work in an appropriate manner within the organization. Additionally, we must become adept at managing expectations that exist within the corporation.

This leads us to question our real purpose in the organization:

What is it your boss wants from you?
What is it you expect from your resources?

The answer to both of these questions is identical. We are often under the impression that our job is to get our work done as quickly and as inexpensively as possible. This turns out to be basically untrue. People have an expectation for the performance of others based upon the work that they have committed to accomplishing. So, if you have committed to complete this project on June 1 of next year, I fully expect that you will follow through with this. Likewise, if one of our resources has committed to completing their task by the end of this week, we fully expect that that will be the case. This leads us to our second truth of management:

2. Do what you said you would do.

This is a very powerful truth. Our leadership does not expect us to get the work done more efficiently or more effectively as much as they expect us to complete the work in the manner that we have promised to complete it. We are expected to take our promises seriously and to make significant personal effort to follow through on those promises. If you think about the resources that report to you, it is likely that you would accept longer schedules and more expensive developments as long as the project team delivered on plan. This is the essential key to project management: we do not need to be concerned about delivering quickly or inexpensively; we need to become dependable to our organization. This ability to generate credibility within the organization will essentially result in the ability to gain the resources necessary to complete our tasks. Without this credibility, it is likely that we will have fewer resources assigned to us, as our supervisors will be less trusting of our appropriate use of those resources. If we instead deliver on our project plan every time, we will likely receive as many resources as we can utilize in our work. This is due to our ability to make our supervisor look good. We often do not consider the fact that when supervisors assign us resources they are risking that these resources will be appropriately utilized for the corporation. Our ability to deliver on promises is essentially equivalent to the efficient use of resources in the eyes of our supervisor. Therefore, the greatest favor you can do for yourself is to complete your project on time and on budget as you promised in your original project plan.

Of course this means significant time involved in the planning phase, and effort made to maintain a specific focus on the work to be accomplished within the project team. Additionally, this requires a great deal of courage on the part of the project manager. If for example, you complete a project plan and find that it will be finished on August 1, you will typically report this to your supervisor and explain that you can accomplish your specification with this budget and this schedule, which will culminate on August 1. Often your supervisor may suggest that the work must be completed earlier, perhaps on June 1. The project manager is required to have the courage to push back on the supervisor, as his or her job is to be sure to have a high probability of success of completing the work that has been promised. Therefore, the project manager must be willing to state that if the project must be done sooner it must be assigned more resources and more money, and the project's scope must be reduced. Without the courage to push back, if we simply acquiesce to managerial requests like this one, we damage our own credibility and our future ability to gain the resources necessary to complete our projects.

When we consider why we get unrealistic demands on our specification, schedule or budget from leadership, we must understand how this relates to the previous discussion. In general, this occurs because we are not building a credible basis for our estimates and proposed work to be accomplished. If we begin to deliver our projects on plan, we will typically find that the resources will no longer be a problem for us. This is due to the credibility we have built up within our organization. If, however, you cannot get the resources necessary, and you receive unrealistic demands from your leadership, it is likely that you have not built credibility based upon your past performance. Therefore, it is incumbent upon good project managers to focus specifically on their ability to do what they said they would do in order to build the credibility that will result in a greatly increased probability of success within the organization. We must build trust, and with trust comes more flexibility and more options for project managers.

This leads us to the third truth of management:

3. Your boss does not want to hear what your boss wants to hear, tell the truth.

This may seem to be a relatively trivial insight, but in fact many project managers do not understand the value of building credibility and trust within the organization. This implies that when you find yourself in a difficult situation and you need to deliver bad news to your supervisor, now is an excellent time to do so. Always tell the truth to the best of your ability within the organizational constraints.

If your project appears to be coming in late or over budget, now is the best time to communicate that to your boss. If your boss is giving you unrealistic goals to complete with regard to schedule or budget coming, now is the best time to go have a conversation with your boss to ensure that you can complete your project with a high degree of probability. Honesty is generally the best policy with regard to projects, and openness will never lead to challenges in your credibility. On the flip side of that coin, however, it is obvious that if we choose to cover up negative information or we refuse to communicate that to our leadership, it is likely that we will damage our long-term trust and ability to complete our projects effectively.

So what is the end result of understanding the various truths of management? It turns out that if we do not adhere effectively to truth number two and truth number three, we are likely to get exactly what we deserve within the organization. To make matters worse, we may actually move heaven and earth to meet unrealistic expectations, which will create further unrealistic expectations. It becomes a vicious cycle that will result in the long-term detriment and failure of most project managers. We need to recognize when we might enter one of these cycles, and instead focus on providing direct feedback that will enable our supervisor to support us and produce a stronger project management performer long-term.

What is the root cause of these problems of unrealistic expectations and cycles of destruction within our organizations? It is best explained with the fourth truth of management:

4. You get what you reward.

Why is this an important insight for us to understand? It turns out that the human very readily understands the concept of reinforcement theory. This series states that the events that immediately follow a behavioral choice will reinforce that behavior. For example, if you choose to touch the stove without considering if it has recently been used, you may burn your finger. This negative response to your behavioral choice may result in you being more careful the next time you go to the stove, checking to see if it is hot before touching it. This concept extends well into our organizations, as people will figure out what produces rewards and they will seek those behaviors in order to maximize their gains.

Unfortunately, we have a long record of reinforcing behavior that is opposite to what we truly desire. For example, consider a department where there are two project managers. One project manager is highly effective and always meets schedules, and works very diligently to assure that promises are kept. The second project manager doesn't get very concerned about missing schedules, and doesn't really make significant efforts in the event that schedules or budgets are slipping. We will also imagine that both of these project managers are fully utilized with the department.

Imagine that you were now assigned a hot new project within the organization. If you are responsible for this department, who gets this new project management assignment? Most of us would select the effective project manager. Unfortunately, from that person's perspective, this is an indirect punishment for doing a good job in the past. He or she is now overloaded and is expected to do extra work because of past successes. We do the same thing to responsible project managers; when many of the projects in our department are overrunning, we steal from the projects that are being responsibly managed in order to cover the sins of those that are not. In the same manner, project managers who do not accomplish a great deal of work are indirectly rewarded for their lack of performance by getting fewer project assignments. We must take great care within our organizations to see that this is not allowed to happen; this requires constant diligence on the part of the project manager.

We have established that it is the responsibility of the project manager to ensure that the work that is planned is completed in an effective manner for the organization. In fact, project managers are the stewards of significant resources for the organization, and therefore must take seriously their promises as outlined in their project plans.

Upon reflection, we realize that once the project plan is complete, the responsibility of the project manager is largely to focus on the people. This focus is essential because the people choose how much effort and productivity to provide for the project. This may be affected by our communication abilities and our diligence with regard to managing the project on a daily basis, but there is no substitute for engaging our project team to ensure success. This is the fifth truth of management:

5. Enable people to make the right choices.

If we spend our time focusing on the individuals working on the project team, it is likely that we will be able to address roadblocks and challenges before they arise. The removal of obstacles and our ability to smooth the path for our project team members result in a greater engagement of the project team, and increase the probability of success of our projects. For these reasons, we considered the truths of management to be critical insights for all project managers:

1. It depends
2. Do what you said you would do
3. Tell the truth
4. You get what you reward
5. Enable people to make the right choices

Understanding these truths will enhance our ability to engage the project team, and will also build credibility and trust within the organization. Knowing and embracing them may seem very trivial and obvious, but very few project management practitioners find themselves able to do so unfailingly. Upon reflection, however, we realize that the project managers we know who have demonstrated great capabilities and have earned great reputations have followed these rules and have adhered to them on a regular basis. Therefore, the truths of management enable us to understand how to build trust and integrity in the organization, and allow us to build coherent and effective project teams.

Understanding People

It is of paramount importance to project managers that they become effective at understanding people. This is most often true during the execution phase of the project, as they must be able to engage their project team, connect with stakeholders, and communicate the status and outcome of the project effectively.

Unfortunately, this also seems to be one of the least understood areas of expertise related to project management. In most project management textbooks, only a few observations are shared with regard to the psychological and sociological aspects of being a successful project manager. This is extremely unfortunate, as it does not clarify for the typical practitioner what areas of interest should be investigated and enhanced.

We will begin by investigating some of the research associated with success in project management. One of the best-known sources for this information is the Standish Group's Chaos report regarding typical performance on projects in the information technology domain. A brief summary of these results is shared in Figure 64.

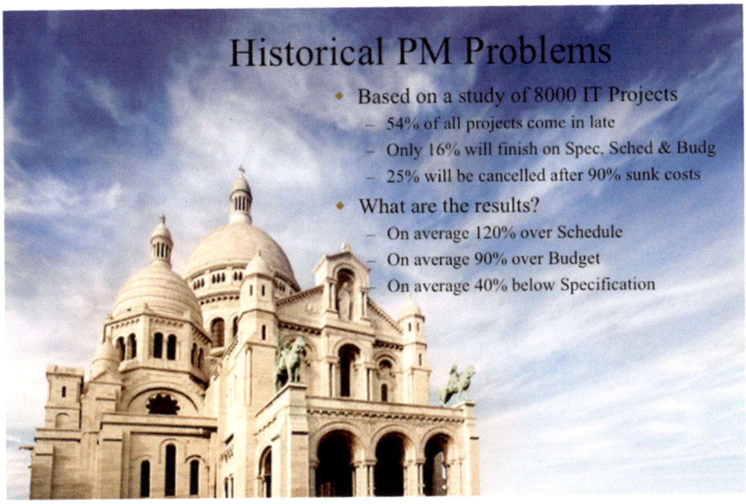

Figure 64: Standish Group CHAOS Results

Foremost, we examine the quality of the study that was undertaken in order to generate these results. The study consisted of eight thousand information-technology projects, self-reported by project managers, with an open-ended format allowing great freedom of interpretation of the results. This leads us to believe several things about this study; first, that it is likely to be a reliable and valid view of a large cross-section of IT project managers in America. This also suggests that we can utilize these results to enhance our own understanding of typical challenges faced by project managers of complex undertakings.

We begin by examining the summary results from this study, the first of which states that fifty-four percent of all projects are delivered late. When this fact is shared with various project management colleagues in my past, it is met with great acceptance. In general, most practitioners suggest that this number is even low, and expect that as many as two-thirds of all projects are likely to come in late in most of our organizations. Of all the projects undertaken, only sixteen percent will meet their original project plan. Thus, eighty-four percent of all projects undertaken are likely to miss their performance goals, often by a very large margin.

On average, the projects that did not meet their schedules were delivered at one hundred twenty percent over schedule. Similarly, the projects that did not meet their original budgets were typically ninety percent over budget. Projects that were forced to give up some of their scope reduced their specification or requirements by an average of forty percent. These numbers are extremely troubling, and we must not forget that they are simply averages. However, this implies that we have many projects that greatly exceed these values in underperforming their original project plan. Apparently we are not alone with our project difficulties.

Due to the fact that we typically budget our departmental use of resources at one hundred percent, it comes as no surprise that mid-way through the performance period, eighty-four percent of our projects are lagging in their project management metrics. If we take proactive action at this point, we must remove funding and resources from some projects in order to shore up the successful completion of other projects. This is extremely unfortunate, as it typically results in our canceling twenty-five percent of all of our projects after we have invested ninety percent of the cost of the project.

Is this good news for project managers? Certainly, this is not good news for organizations, but it does set the bar particularly low for the project manager as a performance metric within the organization. Highly effective project managers would find these values to be disturbing and unacceptable for their project efforts.

We should always understand that there are occasionally insurmountable problems that occur within our organizations, many of which will have negative impacts on our projects, but it is equally important for our project managers to do what they can to meet their commitments. It is the author's belief that the majority of underperforming projects are brought about by project teams who are not engaged and who do not take their promises seriously within the organization.

When the project managers were asked about the causes of project failure, their answers were quite fascinating. The summary of these responses is shown in Figure 65.

More than one-half of the respondents suggested that the project failure was caused by the improper specification of the project's objectives. This response is troublesome, as it illustrates the project manager's belief that this is not the responsibility of the project plan development. Another result stated that bad planning and estimating resulted in nearly one-half of the project failures. This terminology is somewhat troubling as well, as there is an implication that some planning is good and other planning is bad. In reality, some planning is predictive while other planning is less predictive. The difference between the two is largely based on the experience and self-awareness of the project manager responsible for the project planning.

Nearly one-half of the respondents stated that new technology was a confounding factor for their project's success. While this may seem like a valid response, in reality, the information technology department is responsible for new technology introduction in general. This implies that by the very nature of the work, the information technology department should not be expected to meet original project plans, obviously a concerning response.

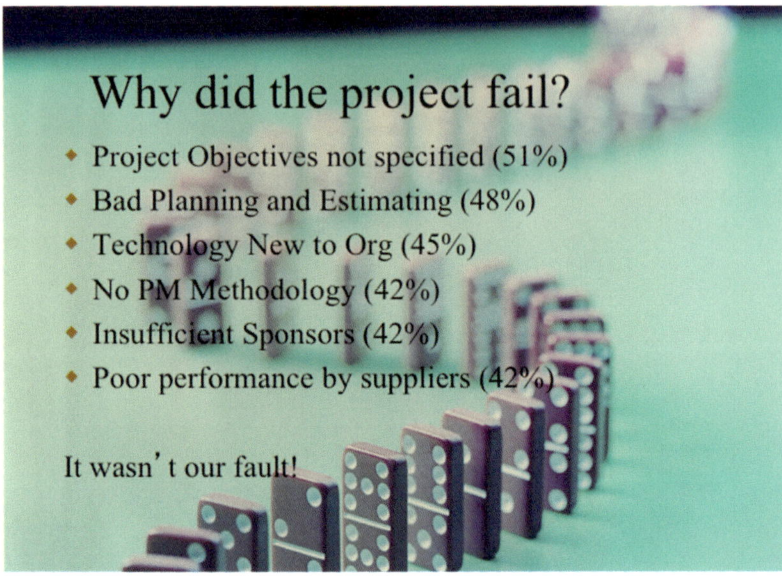

Figure 65: Project Manager Issues

We see a variety of responses in many studies of this type, and these seem to be relatively common excuses given for the lack of predictability during the project execution. Fundamentally, it seems clear that the project managers are suggesting that this was not their fault. In reality, regardless of what occurs either in or outside of your project, the project managers are responsible for the successful completion of the work. Effective project managers would typically respond to a question like this with "there are many mitigating factors resulting in this project performance, but fundamentally I am responsible. This is how I intend to rectify this project performance going forward."

When the study inquired as to developmental areas that would enable them to become more effective in project management, the respondents identified the items specified in Figure 66.

Clearly, the project managers understand fundamentally that they need to increase their project management skills in general. Unfortunately, almost an equal number of respondents noted that they would have benefited from a feasibility study. This would lead us to believe that they were typically following no project management methodology, or they would have undertaken a feasibility study during the project initiation phase of their project.

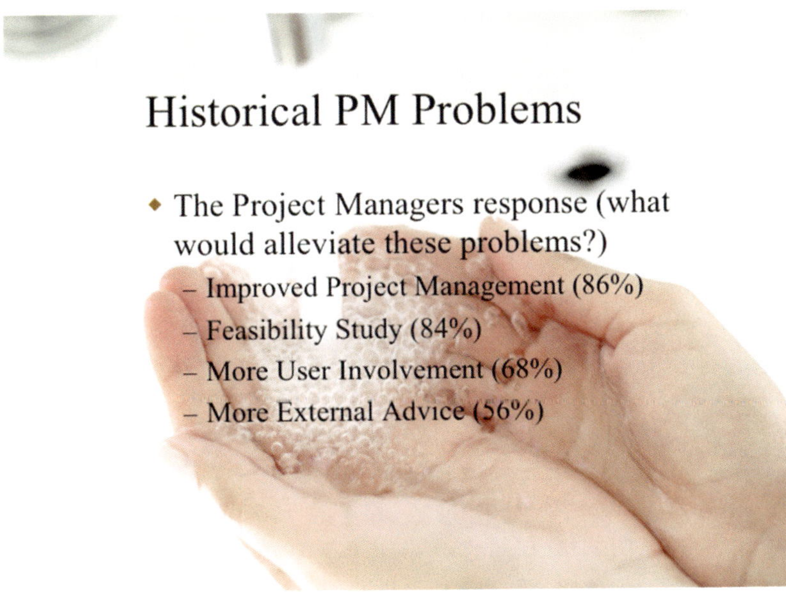

Figure 66: Project Manager Recommendations

As might be expected, there seems to be a lack of personal responsibility associated with these comments and perspectives on what would most benefit the project manager and result in more success in his or her endeavors. It is imperative that project managers understand that they are fundamentally responsible for the effective use of resources within their organizations. They must take it upon themselves to focus on those things that were ineffective in their project performance and develop the areas requiring attention.

Without this personal responsibility, it is unlikely that a project manager can become successful in project management endeavors. In addition to this, without the demonstration of personal responsibility and engagement by the project lead, it is unlikely the project team will engage in the project, thereby reducing the probability of success by a significant factor. Therefore, project managers must undertake project management as a lifelong learning experience, always attempting to improve their approaches and methods in order to produce projects that will be delivered on plan. This requires a significant amount of humility and self-introspection, but will result in significant advances in their performance.

Individual Differences and Personality Traits

In addition to enhancing their own self-awareness, project managers must also become adept at understanding the people who are contributing to the success of their projects. This means that effective project managers will also be effective at relationship building, and are constantly seeking to increase their understanding of the projects' contributors. Fortunately, there are many tools to accomplish this, and we will examine several that are of interest, normally sourced from the psychology domain.

It is common in the psychology domain to attempt to quantify various factors associated with people's behavior and perspectives, in order to enable comparisons and analyses of those various personality traits or individual differences. These can cover a wide spectrum of descriptors regarding individuals, and are typically developed in an attempt to provide a practical taxonomy for any practitioner in the psychology domain. If, for example, you can create a highly reliable and valid instrument to measures a person's creativity, it would be very useful to be able to use that metric to compare various individuals with regard to their creativity.

One of the most common assessments utilized is associated with the general mental or cognitive ability of the individual. We normally associate this assessment with intelligence testing, and this metric will help to determine who we would consider to be smarter or more intelligent. This is a very sophisticated domain and requires specific expertise to be able to implement testing in an appropriate and non-harmful way. Ultimately, it gives us the greatest insight into the potential job performance of individuals, but it can be costly and time-consuming.

Personality testing takes many different forms, and purports to create a taxonomy that enables us to describe an individual's personality in comparison to others. These tests might include the Myers-Briggs type indicator or MBTI, the Minnesota Multiphasic Inventory, social styles inventories, or any of the myriad personality inventories that are currently available in our social media world.

One of the most highly reliable tests is the Myers-Briggs type indicator. Officially, the application of this test requires interpretation by a certified Myers-Briggs consultant, although many abbreviated tests have recently appeared that allow people to gain a quick snapshot into their Myers-Briggs. This descriptor is made up of a four-character sequence describing the propensities of the individual on four different continuums. These include:

Introverted – Extroverted
Intuitive – Sensing
Thinking – Feeling
Judging – Perceiving

The introverted/extroverted scale is based on an assessment of where the individual seeks to gain energy. Introverted people tend to get their energy from within, in spending time on their own and in developing things in an autonomous way. Extroverted people have a tendency to gain their energy from others, are typically more relationship oriented and seek to interact with other people in their development efforts.

Intuitive people are likely to spend a great deal of their time seeking patterns and understanding of complex situations. They reflect on information presented to them before taking action, and can draw conclusions from limited information. Sensing people are more likely to require specific data and direct access to the item of interest, allowing them to gain information directly and tangibly. We can draw a direct correlation between intuitive and right brain thinking, while sensing is more directly related to left brain thinking.

The thinking/feeling scale is associated with the analytical approach to understanding on the thinking side of the scale, whereas the feeling side of the scale is more closely associated with the emotional response to the situation. Thinking would be more closely associated with left-brain thinking, feeling would be more closely associated with right brain thinking.

Judging is a characteristic where people draw conclusions based on the information presented to them, creating judgments or making assumptions where necessary to address shortcomings in their information source. Perceiving people, on the other hand, focus on reading between the lines, and gleaning more information then may be readily apparent from a given situation.

We can begin to better understand the use of a personality inventory by examining a scenario that is designed to place us into a particular mindset prior to beginning. In this example, the reader is urged to read the following situation, placing him or herself into this circumstance prior to taking the subsequent personality inventory. It is important to take a moment and embrace the scenario in order to ensure appropriate results are measured by the inventory.

Scenario 1

You are returning to work after a very enjoyable weekend. As you park your car and make your way towards the front entrance to the organization, you notice the beautiful weather and find you are looking forward to having a good day in the office. When you step inside, you are greeted with a smile from the receptionist. You calmly make your way down the long hallway to your office, third from the end on the left. As you enter, you see the memo you left centered on your desk last Friday, reminding you to check the status of your project and messages in your mailbox this morning in preparation for your status update meeting at 10 AM. As you settle into your comfortable office chair, you reflect on the fact that all of your programs seem to be running effectively and efficiently, and there are no immediate fires to put out. In fact, you are beginning to think that you just might have time for that suggested lunch get-together that was mentioned over the weekend. You reach into your uncluttered desk drawer to find your phone contacts, and begin dialing your friend's number, reflecting on what a good day this is destined to become.

Circle the number that best describes you for each:

Introversion / Extroversion						
Inward	1	2	3	4	5	Outward
Quiet	1	2	3	4	5	Outgoing
Private	1	2	3	4	5	Social
Thoughtful	1	2	3	4	5	Spontaneous
Deep	1	2	3	4	5	Broad
Total of All Five Questions	→	→	→	→	→	
Sensing / Intuitive						
Into details	1	2	3	4	5	Into patterns
Practical	1	2	3	4	5	Innovative
A Realist	1	2	3	4	5	An Idealist
Present focused	1	2	3	4	5	Future focused
Direct	1	2	3	4	5	Creative
Total of All Five Questions	→	→	→	→	→	
Thinking / Feeling						
Objective	1	2	3	4	5	Subjective

Analytic	1	2	3	4	5	Sympathetic
Critical	1	2	3	4	5	Appreciative
Logical	1	2	3	4	5	Personal
Into Data	1	2	3	4	5	Into reactions
Total of All Five Questions	→	→	→	→	→	
Perceiving / Judging						
Flexible	1	2	3	4	5	Firm
Like to Explore	1	2	3	4	5	Like to Decide
Wander	1	2	3	4	5	Structure
Question	1	2	3	4	5	Organize
Casual	1	2	3	4	5	Neat
Total of All Five Questions	→	→	→	→	→	

Now locate your four totals on the following table:

Meyers / Briggs Type						
Introversion	5	10	15	20	25	Extroversion
Sensing	5	10	15	20	25	Intuitive
Thinking	5	10	15	20	25	Feeling
Perceiving	5	10	15	20	25	Judging

Meyers Briggs Types are typically described as four letter sequences of dominant traits. Thus, an individual that scores on the introverted side, intuitive side, thinking side and judging side would describe his or her traits as an INTJ.

Personality in History

Hippocrates described four personality types, which he referred to as humours. Each had associated physical and mental characteristics; the result was a system that was quite subtle in its capacity for describing types of personality.

Humour	Qualities	Element	Personality
Sanguine	Hot, moist	air	Optimistic, red-cheeked, corpulent, irresponsible
Choleric	Hot, dry	fire	Short-tempered, red-haired, thin, ambitious
Phlegmatic	Cold, moist	water	Sluggish, pallid, corpulent, lazy
Melancholic	Cold, dry	earth	Introspective, sallow, thin

In addition, different humors could be combined for more complex personality types: choleric-sanguine, phlegmatic-melancholic, and so on.

Choleric - easily irritated; irascible; inclined to anger; OR angry; indicating anger; excited by anger.

Phlegmatic - hard to rouse to action; specifically, (a) sluggish; dull; apathetic; (b) calm; cool; imperturbable

Melancholic - one affected with a gloomy state of mind; one characterized by excessive gloom, mistrust and depression.

Sanguine – cheerful, optimistic, hopeful. Synonyms = warm, ardent, animated, lively, confident, hopeful.

Transparent Social Styles Inventory

Circle the number that best describes you based on the following scale:

1) Not like me at all
2) Only slightly like me
3) I am somewhat like this
4) I am quite like this
5) Describes me perfectly

I love getting into the details	1	2	3	4	5
I am very conscientious in my work	1	2	3	4	5
I don't like to take risks, I wait until I know	1	2	3	4	5
I don't like having to change my plans	1	2	3	4	5
Once I have my target defined, I stay on track	1	2	3	4	5
I don't like having to make quick decisions	1	2	3	4	5
	Analytic Total				
I tell people like it is, I don't mince words	1	2	3	4	5
I love to get my hands on the work	1	2	3	4	5
I usually make good decisions based on my gut feel	1	2	3	4	5
I really like to be in charge, it makes things happen	1	2	3	4	5
I like trying new things all the time	1	2	3	4	5
I don't like working with the bean	1	2	3	4	5

counters, too restrictive					
	Driver Total				
I love it when my team works together	1	2	3	4	5
I like helping other people get things done	1	2	3	4	5
I am very deliberate in my work	1	2	3	4	5
Sometimes I let rude people get away with it	1	2	3	4	5
I really like things to stay pretty constant	1	2	3	4	5
I would rather agree with others than insist on getting my way	1	2	3	4	5
	Amiable Total				
I love meeting new people, I strike up conversations all the time	1	2	3	4	5
I am really into my work	1	2	3	4	5
If my boss needs me to stay late, it's not a problem at all	1	2	3	4	5
I sometimes end up in a different place than I thought I would	1	2	3	4	5
I don't like the menial work	1	2	3	4	5
I like to think big picture, others do the detail work	1	2	3	4	5
	Expressive Total				

Social Style and Management

According to David W. Merrill, Ph. D., and Roger H. Reid, M.A., "We all say and do things as a result of certain habit patterns, and people make predictions about us because they come to expect us to behave in a particular way - the fact is that even though each of us is unique, we tend to act in fairly consistent, describable ways. All of us use habits that have worked well for us, habits that make us comfortable, and these habits become the social style that others can observe."

The American population is evenly divided among the four social styles: driver, analytic, amiable, and expressive. Each person has a dominant social style, and that style influences the way he works. People will tend to seek out social situations that reinforce their behavior and avoid situations that cause discomfort.

CHARACTERISTICS OF THE FOUR STYLES

Analytical	**Driver**
They live life with consistency according to facts, principles and logic. They often seem to be cool and independent, lacking enthusiasm. They tend to be cautious about extending friendship or showing personal warmth and, initially, will be more concerned with how things get done without need for personal involvement. They seem to act slowly and use time in a deliberate and disciplined manner. They focus on the past to give them direction for the future and prefer to work on a predictable schedule. They look for solid, tangible, practical evidence to support the validity of their decisions. They also require assurance that their decisions will be valid for the future. They appear to be overly concerned with details and organization.	They know what they want, where they are going, and are more oriented toward getting results than pleasing people. They base their decisions on facts and data, will take risks, and prefer to be given options from which they can choose. They are "tell oriented" and voice may sound forceful without speaking loudly. They often speak rapidly. They are fact oriented and do not need feelings or opinions from others. They prefer brief reading material' prefer working alone or directing others. Conversations tend to be brief and even abrupt. They prefer to make their own decisions. They seek power and do not like to be told what to do. They are decisive, pragmatic, and efficient.

Amiable	Expressive
They place a high priority on friendships, cooperative behavior, and being accepted by others. They like to achieve objectives with people using understanding and mutual respect and will accept authority from another person, if the person is friendly and understanding. Talking and socializing is sometimes more important than getting on with the work at hand. They tend to use personal opinions in arriving at decisions and want guarantees of minimal risk; therefore, they frequently stay with the comfortable and known and especially avoid risks that involve personal relationships.	Usually full of energy, they have quick gestures, speaking more rapidly and louder than others. They make decisions quickly, express opinions strongly, and dislike routine. They appear to have more imagination and creative ideas than other people. They are undisciplined in the use of time. They are people who rely on the opinions of others whom they consider important or successful for decision making rather than facts. They wear colorful and sometimes flamboyant clothes and are playful and fun loving. They are outgoing, enthusiastic, persuasive and spontaneous.

People are like thermostats; they are constantly seeking to reach a state of equilibrium or comfort. As soon as another person enters the picture, tension is produced, and each one must reestablish his balance and comfort zone. The challenge for each of us is to determine the proper amount of tension and stress that will provide the proper balance.

Better communication can be achieved when we understand

the other person and treat him the way he wants to be treated. More effective communication and a better relationship can be established by using the following suggestions.

Personality Type Equivalents

Table of Equivalents for the 4 Personality Types				
D.E.S.A.	Dominant	Expressive	Solid	Analytical
Hippocrates Greek Terms (370 BC)	Choleric	Sanguine	Phlegmatic	Melancholy
"What's My Style?" (WMS)	Direct	Spirited	Considerate	Systematic
The P's	Powerful	Popular	Peaceful	Perfect
The S's	Self-propelled	Spirited	Solid	Systematic
The A's	Administrative	Active	Amiable	Analytical
LEAD Test	Leader	Expresser	Dependable	Analyst
ARRAY (Jonathan Knaupp)	Production	Connection	Status Quo	Harmony
Biblical Characters	Paul	Peter	Abraham	Moses
Gary Smalley	Lion	Otter	Golden Retriever (Dog)	Beaver
Geier	Dominance	Influencing	Competence	Steadiness
DiSC(r)	Dominance	Influencing of Others	Steadiness	Cautiousness/ Compliance
McCarthy/4MAT System	Common Sense	Dynamic	Innovative	Analytic
Merrill / Wilson	Driver	Expressive	Amiable	Analytic
Plato (340 BC)	Guardian	Artisan	Philosopher	Scientist

Kretschner (1920)	Melancholic	Hypomanic	Anesthetic	Hyperasthetic
Sprangler (1930)	Religious	Aesthetic	Theoretic	Economic
Fromm (1947)	Hoarding	Exploiting	Receptive	Marketing
Psycho-Geometrics (1978)	Triangle	Squiggle	Circle	Square/Rectangle
Type A or B	Type B Motivated	Type B Messy	Type A Casual	Type A Compulsive
PSI	Controller	Promoter	Supporter	Analyst
Broken leg Reclaiming Youth at Risk	Mastery Achiever Power	Belonging Attached Significance	Generosity Altruistic Virtue	Independence Autonomous Competence
Enneagram	Adventurer Achiever	Helper Romantic	Peacemaker Observer	Asserter Perfectionist
Animals	Bear	Monkey	Dolphin	Owl
True Colors(r) (1978)	Green	Orange	Blue	Gold
Children's Literature	Rabbit	Tigger	Pooh	Eeyore
Charlie Brown Characters	Lucy	Snoopy	Charlie Brown	Linus
Jane Austen Novel Characters	Emma Woodhouse	Lydia Bennet	Elizabeth Bennet	Marianne Dashwood
Comics	Jason	Snoopy	Cathy	Ziggy
Who Moved My Cheese? (by Spencer Johnson, M.D.)	Sniff	Scurry	Haw	Hem
The Celestine Prophecy (by James Redfield)	Intimidator	Poor Me	Aloof	Interrogator

Sources:
http://www.ylcf.org/you/personality-equivalents.htm
http://macdl.org/edu/socmgmnt.htm

Scenario 2

You pull into the parking lot under overcast skies. It seems like you exited this lot only a short time ago, and when you glance down at your watch you realize it had been less than twelve hours earlier. You have been working a backbreaking schedule in order to meet delivery on your major project effort. As you enter the office, you notice that there is no receptionist on duty, and you make your way down the long hallway carrying your ever-increasing workload in your briefcase. The weekend was not enjoyable; in fact, it was nothing short of a disaster. It started out well at first, but then the phone rang and your inept boss insisted that you work on Saturday to clean up a mess caused by a lazy colleague. This is the third Saturday in a row you have been called in, and you have had enough of it. You are also beginning to sense trouble at home, even though you spend so little time there. As you enter your office, you notice that no one has broken in and completed all of your work, and the daunting stack of paperwork you left yesterday remains in the center of your desktop. As you set down your things, you notice a new memo on the top of the pile. You crane your neck slightly to read it as you reach over to turn on your computer. It announces that management has elected not to award profit-sharing checks this year, even though the organization has shown a profit. In fact, the organization requires you to take a five percent decrease in salary over a period of six months in order to meet the forecasted goals of the company. You find yourself grinding your teeth, thinking about the brand-new Learjet the CEO just purchased. You sit down heavily in your uncomfortable chair, and reach for the phone as it rings with the first of the day's long series of problems.

Circle the number that best describes you for each:

Introversion / Extroversion						
Like to Do	1	2	3	4	5	Like to Think
Performing	1	2	3	4	5	Observing
Discuss a Book	1	2	3	4	5	Read a Book
Wild	1	2	3	4	5	Domestic
Like a Dog	1	2	3	4	5	Like a Cat
Total of All Five Questions	→	→	→	→	→	
Sensing / Intuitive						
Like the big picture	1	2	3	4	5	Like to fine-tune
Creative	1	2	3	4	5	Pragmatic
Idealism	1	2	3	4	5	Realism
Long-term focus	1	2	3	4	5	Plan it out
Artistic	1	2	3	4	5	Detailed
Total of All Five Questions	→	→	→	→	→	
Thinking / Feeling						
Free-thinking	1	2	3	4	5	Rational
Empathetic	1	2	3	4	5	Practical
Emotional	1	2	3	4	5	Realistic
Focus on people	1	2	3	4	5	Focus on ideas

Into Discussion	1	2	3	4	5	Into Conclusions
Total of All Five Questions	→	→	→	→	→	
Perceiving / Judging						
Confident in self	1	2	3	4	5	Open to ideas
Like to define	1	2	3	4	5	Like to consider
Plan each detail	1	2	3	4	5	Travel without plans
Go the right direction	1	2	3	4	5	Go with the flow
Established	1	2	3	4	5	Independen t
Total of All Five Questions	→	→	→	→	→	

Now locate your totals in the following table:

Meyers / Briggs Type						
Introversion	25	20	15	10	5	Extroversion
Sensing	25	20	15	10	5	Intuitive
Thinking	25	20	15	10	5	Feeling
Perceiving	25	20	15	10	5	Judging

The reader will note that they have completed two different Myers-Briggs type indicator inventories. These were based on wholly different scenarios, and will result in significantly different results for the participants. This is troubling for people reflecting upon the solidity and reliability of the personality inventory, as they expect that their results would not vary significantly. However, the resulting personality measured by any given inventory is often impacted dramatically by the mindsets of the individuals undertaking the test. This is why the scenario was presented prior to the inventory itself; it actually illustrates three very important types of personality traits possessed by all of us:

Intrinsic personality traits are hardwired in our brains from birth. These are the traits with which we are most comfortable, and that we rely on when in challenging situations. They will largely govern our behavior during our developmental years, and as we progress through our maturity we develop additional personality traits. This is very useful for us to understand, as it implies that those individuals who were not blessed with the exceptional personality required to accomplish a given job may in fact be able to develop those skills by developing alternative personality traits.

Learned personality traits are those that we have recognized would be beneficial for us. For example, if you are a strong introvert on your personality inventory, but you would also like to be a theater performer, you must develop your extroversion in order to become successful. It turns out that humans are remarkably effective at developing alternative personality traits, although exercising them requires much more energy than relying on our intrinsic traits. Therefore, if you are a strong introvert who is operating in an extroverted position, you will likely be more physically and mentally exhausted than an individual who is operating within their intrinsic traits.

The third type of personality traits is known as **latent** traits. These are also hardwired into our systems, but typically only exhibit themselves in highly traumatic situations. Latent traits explain why we observe soft-spoken individuals becoming heroic at times of great need, or individuals exhibiting new personality traits in an emergency situation. These traits may be useful for us to understand, but they are typically only exhibited when we trigger our fight or flight response. Hopefully, this is seldom triggered in your project management efforts, but nonetheless it is important to be aware of this type.

Another method of measuring personality traits has been developed largely within the psychology domain, and is known as the Big-Five Model of personality traits. This model was developed by examining all of the English words in the dictionary, identifying those that could be construed as descriptive of a potential personality trait. The original list constituted 17,953 different words. It was pared down utilizing various techniques until the researchers had identified thirty-five. Once factor analysis was applied, they had identified just five unique personality traits. This is noteworthy, because we recognize that since the time of Hippocrates most personality inventories have been conducted utilizing four individual traits. This suggests that we have a more complete understanding by the utilization of our entire language as a basis for the study. It is noteworthy to mention that when conducted in other languages, the same study has also discovered five independent personality traits.

The Big-Five Model

Openness to Experience

Openness to experience is a measure of risk propensity in the individual. People who score highly in this are interested in new experiences and new opportunities. This will usually indicate good leadership potential.

Conscientiousness

Conscientiousness is associated with the amount of rigor we bring to our efforts. Highly conscientious people are likely to do a very detailed and complete job, and are expected to have a higher degree of accuracy in their work. Low conscientiousness is typically associated with laziness or shoddy workmanship.

Extroversion

The extroversion trait is very similar to that identified in the Myers-Briggs type indicator. This is a measure of where the individual chooses to source his or her energy, from others or from within. People exhibiting a high degree of gregariousness will typically score high in this trait.

Agreeableness

Agreeableness is a trait that is unique to the Big-Five Model, and does not typically appear in the personality types previously described. It relates to the propensity of the individual to achieve consensus as opposed to allowing conflict. This conflict avoidance will drive unique behaviors that will result in more significant relationships and an easier communication function for the individual.

Neuroticism / Emotional Stability

Emotional stability is a measure of centeredness or stability. People with high emotional stability are unlikely to be rattled by day-to-day challenges or victories within the organization; they typically do not get very high and do not get very low with regard to events on their project. This trait is also typically associated with highly effective project managers, as there is a tendency to not allow the day-to-day noise of the project to rise to undue levels of importance.

The following inventory is known as a transparent bipolar inventory, similar to those previously taken for personality tests. The inventory is considered transparent when the individual taking the inventory can clearly see what is being measured. Bipolar refers to the dichotomous terms that are used to create a continuum for the test-taker to self-score. The results of this type of test will be very similar to those of the Myers-Briggs type indicator, and may provide some insight to the individual about themselves or others who have taken the test.

Please indicate your preferences for the following pairings of words in the space provided by circling the appropriate number. Consider yourself to be acting in the role of Project manager.

	Very	Moderately		Neither		Moderately		Very		
			Introversion -Extroversion							
introverted	1	2	3	4	5	6	7	8	9	extraverted
unenergetic	1	2	3	4	5	6	7	8	9	energetic
silent	1	2	3	4	5	6	7	8	9	talkative
timid	1	2	3	4	5	6	7	8	9	bold
inactive	1	2	3	4	5	6	7	8	9	active
unassertive	1	2	3	4	5	6	7	8	9	assertive
unadventurous	1	2	3	4	5	6	7	8	9	adventurous
			Pleasantness or Agreeableness							
cold	1	2	3	4	5	6	7	8	9	warm
unkind	1	2	3	4	5	6	7	8	9	kind
uncooperative	1	2	3	4	5	6	7	8	9	operative
selfish	1	2	3	4	5	6	7	8	9	unselfish
disagreeable	1	2	3	4	5	6	7	8	9	agreeable
distrustful	1	2	3	4	5	6	7	8	9	trustful
stingy	1	2	3	4	5	6	7	8	9	generous
			Conscientiousness or Dependability							
disorganized	1	2	3	4	5	6	7	8	9	organized
irresponsible	1	2	3	4	5	6	7	8	9	responsible
negligent	1	2	3	4	5	6	7	8	9	conscientious
impractical	1	2	3	4	5	6	7	8	9	practical
careless	1	2	3	4	5	6	7	8	9	thorough
lazy	1	2	3	4	5	6	7	8	9	hardworking
extravagant	1	2	3	4	5	6	7	8	9	thrifty
			Emotional Stability							
angry	1	2	3	4	5	6	7	8	9	calm
tense	1	2	3	4	5	6	7	8	9	relaxed
nervous	1	2	3	4	5	6	7	8	9	at ease
envious	1	2	3	4	5	6	7	8	9	not envious
unstable	1	2	3	4	5	6	7	8	9	stable
discontented	1	2	3	4	5	6	7	8	9	contented
emotional	1	2	3	4	5	6	7	8	9	unemotional
			Intellect or Sophistication							
unintelligent	1	2	3	4	5	6	7	8	9	intelligent
unanalytical	1	2	3	4	5	6	7	8	9	analytical
unreflective	1	2	3	4	5	6	7	8	9	reflective
uninquisitive	1	2	3	4	5	6	7	8	9	curious
unimaginative	1	2	3	4	5	6	7	8	9	imaginative
uncreative	1	2	3	4	5	6	7	8	9	creative
unsophisticated	1	2	3	4	5	6	7	8	9	sophisticated

Relating the Bi-polar Inventory to your Meyers-Briggs

Trait	*Descriptor*	*Trait*
Low Extroversion *Introversion*	*Big-Five* *MBTI*	*High Extroversion* *Extroversion*
Low Intellect *Sensing*	*Big-Five* *MBTI*	*High Intellect* *Intuition*
Low Agreeableness *Thinking*	*Big-Five* *MBTI*	*High Agreeableness* *Feeling*
Low Conscientiousness *Perceiving*	*Big-Five* *MBTI*	*High Conscientiousness* *Judging*

There are many other personality inventories, which are available for individuals to utilize at their discretion, any one of which may be useful in helping to gain insight into your own particular propensities as well as others'. Taking the time to understand other individuals will result in a heightened ability to communicate effectively with them and engage them in your project efforts. Some of these other personality tests include:

Locus of Control – master or pawn

Authoritarianism – either like and exploit the authority, or don't

Machiavellianism – ability to gain and use power

Self-Monitoring – can fit into the environment quickly and seamlessly

Risk Propensity – can make decisions with less information

The Need For Communications

Why do we care?

"This would be a great place to work if it weren't for the people."

-- Anonymous

There are typically two sides to any manager's job. The first deals specifically with the day-to-day task requirements of the effort. The second deals specifically with the people, and has to do with how people feel they are treated during the performance of whatever tasks are being accomplished. This dimension of the manager's job requires understanding of one's self as well as an understanding of others. Without this it will be difficult, if not impossible to communicate effectively with people, to build mutual trust, and to motivate and influence employees.

For example, have you ever wondered:

Why someone took offense at some innocent remark?

Why your thoughts are sometimes so different from your feelings?

Why some people can't or won't accept sound advice?

Why your actions sometimes differ from your stated beliefs?

Why you can't always find a rational explanation for someone else's behavior?

Why your interactions with some people always seem to leave you with negative feelings?

What's going on here?

The purpose of this reflection is to introduce you to simple and successful methods for improving your interpersonal communication skills. Whether or not we consciously think about it, communication is always purposeful, as is any other human behavior. We use communications for many different reasons; to exchange information, to let our desires be known, to respond to others, and for many other transactions. We also use communications to impact and control the emotional connection between others and ourselves.

For example, assume you meet a co-worker in the hallway and the following exchange takes place:

You: "Hello, Tom, I see you're back from vacation."
Tom: "Yes, we had a wonderful time. I took five rolls of pictures."
You: "Great, have a nice day."

You have indulged in a communication ritual. Your first comment recognizes Tom and his perspective, and your last comment maintains or increases the emotional distance between you. You could have responded to Tom's remark about taking pictures with a statement related to an interest in seeing the photographs. This would undoubtedly have decreased the emotional distance between you, but human beings both consciously and unconsciously control the closeness of others, primarily using communication.

Above all, communication is used to transfer information. What most people do not realize is that different kinds of data move with different speeds through most organizations, and can be significantly biased and tempered by both the manner and medium with which it is communicated.

We must hone in our communicative skills in order to better accomplish our work and more effectively influence people within the organization. This requires significant knowledge in a variety of domains relating to psychology, sociology, logic, and organizational effectiveness.

To be more effective, it is clear we must be active at communication:

We need to be clear about why we are communicating.

We need to know to whom we wish to communicate as well as to whom we should be communicating.

We need to know the best manner in which to communicate.

We need to know the best time and place to communicate.

We must select effective modes of communication.

What Are Communications?

Communication Theory

Communication is focused on understanding how to more effectively interconnect with our colleagues, the mechanisms we use to do so, and how we they might be improved. The most common communication process diagram is shown in Figure 67.

Sender –	The party sending the message to another party.
Encoding –	The process of putting thought into symbolic form.
Message –	The set of symbols the sender transmits.
Media –	The communication channels through which the message moves from sender to receiver.
Decoding –	The process by which the receiver assigns meaning to the symbols transmitted by the sender.
Receiver –	The party receiving the message sent by another party (also called audience or destination).
Response –	The set of reactions that the receiver has after being exposed to the message.
Feedback –	The part of the receiver's response that the receiver communicates back to the sender.
Noise -	Unplanned static or distortion during the communication process.

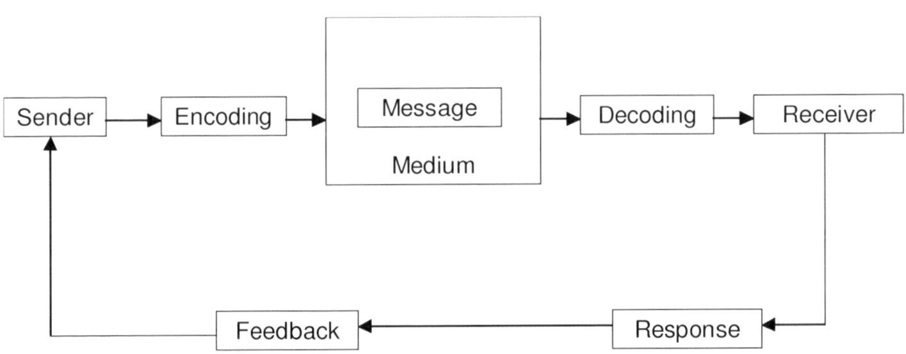

Figure 67: The Communication Process

Methods of Communication

Our methods of communication are highly complex, and require some consideration. We know that successful communications occur when the receiver understands the intent of the sender within the context of the message. Unsuccessful communications occur when the receiver does not understand the intent of the sender, and takes subsequent actions on that incorrect understanding.

Oral communication is one of the most common methods that we utilize. It also happens to be highly effective, as it is typically a very personal exchange and results in a two-way exchange of information. In general, it provides an immediate feedback loop for the sender and can generate insights and learning on both sides.

In addition to the oral communication content, there are a variety of body language factors that also play a very important role. They include:

Body language (kinesics)

1. Eye contact
2. Posture
3. Facial expressions
4. Arm and leg positions
5. Body movement

Many of these body language factors will also be culturally based. As an example, in America it is considered polite to make direct eye contact with those with whom you are communicating. However, in China or Japan one might be considered extremely rude for doing exactly the same thing. Therefore, we must be not only aware of the commonly accepted body language signals for our organization and culture, but also those interpreted by foreign cultures.

Additionally, we might consider the impact of space on our exchanges with other individuals. It turns out that people are highly aware of the physical separation between parties during an interaction, and the use of space can be a major factor in the communication process. For the American culture, this use of space includes:

Use of space (Proxemics)
Personal

a. Intimate (one to eighteen inches)
b. Personal (eighteen inches to four feet)
c. Social (four feet to twelve feet)
d. Public (over twelve feet)

Therefore, when taken in its entirety, communications can be a highly complex and challenging subject to understand for the project manager. To gain an appreciation for its importance, we must consider that research suggests that when we communicate with another individual our body language constitutes fifty-five percent of our message. The tone of voice we utilize and the tone of our message will constitute another thirty-eight percent. *Our words only constitute seven percent of our communication content*. Without a good understanding of these factors, is it any surprise that many project managers have great difficulty in effectively interacting with the stakeholders on their projects? It is well worthwhile for the project manager to gain insights in this area with the intent to improve their communications throughout their organizations.

Written communications are equally as complex, and are

impacted by things like writing style, grammar and lexicon, efficiency and the intellectual level of word choice. It is imperative for project managers to be not only precise, but also rigorous in their attempts to ensure that miscommunication does not regularly occur on their projects. While this is a laudable goal, it is unlikely that all may be eliminated. To maximize effectiveness, it is recommended that project managers develop communications plans early in the life cycles of their projects.

The Project Communications Plan

Project Management communications are crucial as they create the link between the project and the organization. To effectively communicate in written form requires a systematic process. This will provide the critical links among the people, ideas and information that are necessary for Project Management success. Everyone involved in the project must be prepared to send and receive communications, and must understand how they impact the project as a whole.

There are typically three processes required to effectively communicate project status. They include:

Communications planning
Information distribution
Performance reporting

These processes will interact with each other and in other areas of the project, as well. Although we may discuss the processes as discrete elements with well-defined interfaces, in practice they overlap and interact in ways that cannot always be understood or easily detailed.

Project communications will differ significantly from standard communication models. Communication models include issues of sender -- receiver models, feedback loops, and barriers, as well as the choice of media and the selection of how to communicate. Writing style, which might include active versus passive voice, sentence structure, choice of wording and the actual presentation of this information, including body language and design of presentation aids, will all impact the content and acceptance of the communication transaction. These issues extend to beyond the scope of this chapter. We will focus upon the formalized communications associated with projects.

Communications Planning

Communications planning involves determining the information and communications needs of all stakeholders, including what information, when it will be needed, how it will be given, and by whom. Identifying the information needs of the stakeholders and determining a suitable means of meeting those needs is an important factor for project success.

Communications requirements are the sum of the information requirements of the project stakeholders. They might include:

Project organization and stakeholders' possible relationships
Disciplines, departments, and specialties
Logistics of how many individuals will be involved with the project and at which locations and at which times
External informational needs

Information Distribution

Information distribution involves making needed data available to the project stakeholders in a timely manner. It includes implementing the communications plan, as well as responding to unexpected requests for information. These communications may include work results and revisions to the project plan.

Several issues will play a significant role in the effectiveness of information distribution. These include communications skills:

Written and oral, listening and speaking
Internal to the project and external to the project
Formal, in the form of reports and briefings
Informal, such as memos and conversations
Vertical communications and horizontal communications

Information Retrieval Systems

Team members and stakeholders may share information through a variety of methods; including most effectively, electronic archives, project management software, and database systems that allow access to the technical documentation of the project. Project information may be distributed using a variety of methods including project meetings, hardcopy document distribution, shared access to network electronic databases, fax, e-mail, voicemail, videoconferencing, and project updates.

Fortunately for the project manager, the use of the effective information distribution creates a valuable project record.

This can later be used in both project reporting and in presentations related to the project. In fact, most project managers will find the project records invaluable in assessing the post-mortem analysis of the project.

Performance Reporting

Performance reporting involves collecting and disseminating performance information to provide stakeholders with information about the status of the project. This process includes:

Status reporting, that will describe where the project stands.

Progress reporting, that will describe what the project team has accomplished during the previous period.

Forecasting: predicting the outcome of the project and expected progress

Performance reporting should generally provide accurate information on specification, schedule, and budget. Some projects will also require incidental information such as risk management and procurement status.

Many tools and techniques exist for accurate performance reporting. These tools may include performance reviews, variance analysis, trend analysis, and most importantly, earned value analysis. The effective use of earned value analysis will provide the project manager with an accurate assessment of the status of the project with respect to both the efficiency of the project team, as well as the spending rate of the project.

Business Rhythm

Business Rhythm is the term coined to establish a regular set of recurring meetings that are defined in a manner that assures updates cascade (upward) throughout the organization. The primary goal of establishing an effective business rhythm is to ensure that the information presented at periodic updates is both relevant and fresh.

In many large organizations, considerable time is spent in generating and providing updates to higher levels. In fact, if their true cost were calculated, many leaders would rethink the policy of demanding regular and detailed updates. In any case, the use of a business rhythm can dramatically reduce the workload on the associates providing them by establishing a tiered method of updating for all individuals.

As an example, to establish a business rhythm within a large manufacturing company, each of the associates within the company is provided the opportunity to attend a weekly staff meeting. At the weekly staff meeting, employees have the opportunity to update their managers on the current statuses and outlooks of their projects and tasks. A second meeting is scheduled for the First-line Managers following the individual staff meetings. A monthly meeting is established for the Senior Manager level, where all of the functional projects are updated based upon information received at the lower levels. This monthly meeting is scheduled to occur just prior to the quarterly update meetings that occur throughout the year.

Many different constructions can be defined for an appropriate business rhythm, and consideration should be given to the reduction of materials that are generated and presented numerous times within the organization. By staggering the updates from the bottom of the organization to the top, many materials can be consistently reused for higher-level status updates. An example business rhythm schedule is shown in Figure 68.

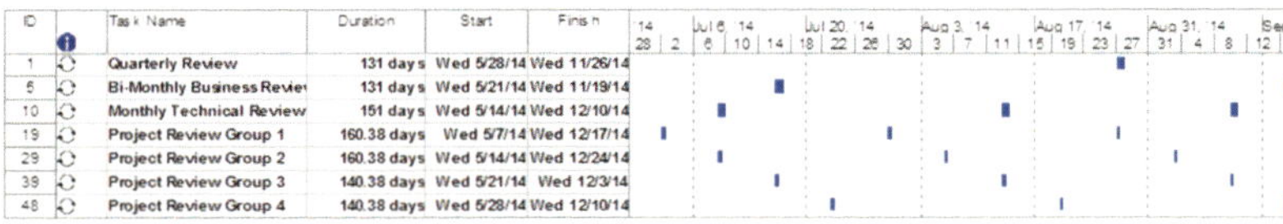

Figure 68: Business Rhythm Schedule

Project Meetings and Status

Meetings should be planned periodically to ascertain status and predict outcome.

1. Set an agenda for project meetings to review progress
2. Define action points against each item on the agenda
3. Review the items on the critical path
4. Report if the cost or time limit exceed plan
5. Report progress at the end of each stage of the project
6. Monitor issues that may be causing concern

Project Reviews

Project Status will provide timely information on the current status of the project, but the use of Project Reviews is recommended to assure the project achieves all objectives and satisfies the customer.

Project Review Key Objectives:

Technical review of system and / or sub-systems to confirm they meet Functionality, Robustness, Specifications and Customer Needs
Cross-disciplinary review to ensure optimized *system* solution
Documentation to comply with Product Development Process and support Design History File

Results of conducting Project Reviews:

Catch mistakes at the design stage, different eyes looking at it
Save money on debug, iterations, expedite fees (payments)
Product meets specifications and customer needs
Team coherence on design and a better product

Participants for a Successful Review:

Who should be included in design reviews and does it depend on the stage of development?
All stakeholders in the project as a starting point
Production
Marketing (represents Customer)
Regulatory / Quality
Project / Program management
Always invite all representatives. Require attendance based on stage of development and topic of review. The team owns the design and is responsible for successful completion, so critiques are of value and should be encouraged.

Running a Successful Review:

Materials – distribute them before the review

Relevant Specifications and Requirements

Detailed design drawings, schematics, simulations, flow charts, test plans, etcetera

Known issues, discrepancies with specifications, conflicts with other disciplines

Agenda with timeline by discipline

Structure – review the project in its entirety by discipline and subsystem

Start on time and stay on schedule (Project or Engineering Manager leads)

Not a formal presentation, rather a discussion approach, so encourage interruptions

Focus on improving the design, no personal attacks

Action items defined for changes, items to check, functionality changes, etc.

Decisions about design approach, specification changes, trade-offs, etc. should be made

Meeting minutes including attendance, references to materials covered, reference to actions / issues / decisions list

Need documented closure on all actions before completing the next design step, as this becomes a gate

Add to Design History File (including copies of materials reviewed)

Concept Design Review (Preliminary Design Review)

Concept review / update

System level architecture is the focus

Identification of high-risk areas of design with focus on technology challenges

Modeling / simulations / breadboard results demonstrating risk mitigation for high risk areas of design
Goal: To provide approval to begin detailed design work

Participants
All appropriate team members
Team members from related projects in the program
Balance group size with required participation

Materials
Hardcopy of agenda, all drawings, schematics, specs, modeling and simulation results
Physical prototypes if available
All materials distributed prior to the meeting

Review Execution
One - two hours (good planning to optimize resource time commitment)
Can break into multiple reviews distributed over days / weeks
Clearly stated goals and objectives with list of materials to be reviewed
Examine if the feasibility is appropriate, and if the approach (with data to support) makes sense to those involved
Be prepared to articulate risks and mitigations based on proof of concept prototype activities (linked to specifications, traceability)
Project manager leads review
Meeting notes required with Actions / Issues / Decisions
All materials and notes to Design History File

Detailed Design Review (Critical Design Review)
Participants: Same as previous review
Materials: Same as previous review plus detailed work accomplished
Goal: To provide approval for design release for production

Review Execution
Two+ hours (good planning to optimize resource time commitment)
Clearly stated goals and objectives with list of materials to be reviewed
Highly technical, quantitative review of **all** aspects of design, all design drawings, all software, etc.
Be prepared to articulate detailed analysis of designs, decisions, and trade-offs
Review of specifications (line-by-line) to ensure design completion and adherence
Completed review should give everyone the feeling that this is an imminent success and release for build

The Art of Delegation

A manager has a wide variety of duties:
To decide what needs to be done;
To tell someone to do it;
To listen to reasons why it can't be done;
To follow up to see if in fact it has been done;
To discover that has not been done;
To inquire why the manager's wishes were not followed;
To listen to excuses from the person who was assigned to tasks;
To follow up again to see up the thing has been done;
Only to discover that it has been done incorrectly;
To describe how it should have been done;
To agree that as long as it has been done it might as well be left in its present state;
To wonder where to find effective personnel who can do things correctly;
To consider that any probable successor would be just as bad or possibly worse;
To reflect sadly that one could have done it right in twenty minutes, and as things turned out, one has had to spend two days to find out why it has taken three weeks for someone else to do it wrong.

-- Anonymous

Delegation

To delegate properly, a manager must have a sense of perspective.

The effective manager must remember the 80 -- 20 rule:

Eight percent of all management decisions do not need to be made.

Managers earn their money on twenty percent of their decisions. Managers must identify which decisions are which. They must effectively have a high batting average.

There are many decisions that can be defined as either a low -- medium -- or a high priority task, and the effective manager must recognize which of these decisions are appropriately delegated. Then there is the class of decisions that would be loosely described as "you bet your job" type decisions.

Getting Tasks done through others

There are typically three methods of work distribution that are commonly followed within the organization.

The first of these methods is **assignment**. The assignment is characteristic of detailed work typically annotated by a specific procedure or schedule that might require a high degree of supervision. This is the method of management where the manager tells everything to the worker.

The characteristics of **abdication** are such that the manager cares little or nothing about the work that must be completed. In this case the manager simply ignores that the task be completed. This abdication might be accomplished by doing nothing, or by delegating the tasks to an individual who was not prepared to complete the task effectively.

Delegation is the method where the manager communicates the desired outcome, the appropriate schedule, and all known constraints to the individual intended to accomplish the work. This is the most effective way of accomplishing work distribution within the organization.

Why Delegate?

Delegate to save time. Delegating appropriately can save you two to fourteen hours a week, or fifteen days per year.

Delegate in order to focus more on high-yield activities.

Delegate in order to participate in activities that will upgrade your skills, which in turn will be helpful both to you and to the organization.

Delegate to motivate your team and enable it to surpass itself, by increasing its responsibilities and allowing it to break out of its usual work routines. Example: a task you consider routine could prove an interesting challenge to one of your subordinates.

Delegate to assess employee productivity and potential by specifying performance standards, realigning objectives, clarifying the means to achieve them, and acting as a mentor.

Delegating provides your colleagues and employees with an opportunity for training, a source of motivation, a means to develop team spirit, and a useful tool to consolidate team synergy.

Obstacles to Delegating

Quality: A delegate believes he or she can do a better job himself or herself, perhaps due to insecurity or a lack of self-confidence. This behavior will instill a defeatist attitude in anyone who works with you.

To minimize this psychological barrier, you must delegate simple tasks and clearly explain what the task involves, while defining standards and outlining procedures in a clear manner

Time: Some argue as follows: "I'd need as much time to delegate this task as to do it myself!"
Solution: Learn to distinguish repetitive tasks from occasional tasks, to delegate repetitive tasks, and to delegate occasional tasks when you can.

Ego: Some think, "If I delegate, people will think I'm lazy or incompetent."
Solution: Make people understand that your role is to manage, not to do all the work yourself.

Encourage those around you to assume more responsibility, delegate as much as possible and keep track of work that has been delegated.

Lack of self-confidence: Some think, "If I delegate too much, someone will take over my job." Solution: draw up a list of realistic goals according to the time available.

Meet with your hierarchical superior and discuss these goals to obtain approval of any changes you wish to make.

Delegate everything you can, in order to save time for (new) high-yield activities.

Fear of others' reactions: Some feel, "People will react negatively."
Solution: Meet with the people you work with and learn what their interests are, explain your intentions clearly, experiment with your team and seek out the reasons for lack of interest, and find solutions to problems you encounter.

Types of Delegation

Horizontal delegation = A manager delegates a task to another manager at the same level.

Vertical delegation = A manager delegates an activity, task or group of tasks to their subordinate.

Reverse delegation = An employee delegates a task to a superior. How does this process work?

1. The boss delegates the activity.
2. The employee consults the boss, since there seems to be a problem.

3. The boss (to be nice) takes charge of the problem and says he or she will look after it; hence the role reversal.
4. Thus, the employee controls the boss. This situation occurs frequently.

Solution: Become aware of the reverse delegation phenomenon, otherwise you will waste time, promote negative attitudes, and be a poor mentor. You must listen carefully, review the situation with the employee, spell out your expectations once again, help the employee (not do the work for him or her) and encourage the employee to find other solutions.

How to delegate

The "secret" of effective delegation is to define:

The task
The accountability requirement
The responsibility being given
The authority being given

Definition of the task: The task should constitute a "whole," that is, a project for the individual to undertake. It should be worthwhile and relate to the overall mission of the organization. The purpose of delegation is two-fold: to accomplish the task and to develop the employee. Both are important.

Some aspects of good task definition are:
Clearly defined outcome(s), what is the deliverable?

Clear statement of the time frame, when is the deliverable due?
Specification of the monitoring procedures to be used; when will review meetings be scheduled? When are interim reports due?
An indication of how the outcome will be used; will the result be implemented immediately or will it be sent for higher level? Is it to be used as input to a decision; if so, when will the decision be made?

Accountability Requirement: The manager should specify the task parameters so that the delegate realizes the limits of his or her authority and will not make unilateral decisions prejudicial to the organization. The delegate should also learn the manager's expectations for periodic monitoring. The monitoring itself may be delegated; for example the individual may be asked to produce regular written reports.

Responsibility Delegation: The manager remains accountable for the outcome(s) of the task but must confer on the delegate the power to make decisions needed to ensure success of the project. Once the boundaries of responsibility are clearly defined, the delegate should be allowed to do the work without any interference. In delegating a task, the following should be made clear (preferably in writing):

The overall goal (of which the task being delegated is a part.)
The precise outcome(s) required;
How far the delegate can interpret events and the extent to which his or her actions must conform to set rules;
The nature of circumstances that will require authorization by a higher-level person how (and at what intervals) accountability monitoring will take place.

Authority given: A frequent complaint made against managers is that they give responsibility without authority; that is, they ask an employee to undertake a task without giving him or her the necessary resources to succeed. At the initiation of the delegation, the manager should transfer to the delegate the following:

The financial resources needed:
A major task (think of it as a project) should carry its own budget and the delegate should have the power to spend money as required by the task without requesting permission from a superior. Because financial accountability remains with the manager, it may be necessary to have periodic reviews of the financial situation to be sure that funds are being spent appropriately and that the amount allocated was sufficient for the task. Typically the reviews take the form of an accounting report submitted on a regular basis.

The material resources needed, e.g., materials, facilities, or technical aids:
If resources are to be shared with others, the priority of the particular project should be clarified at the outset. Is the delegate to wait in a queue for a shared resource? If so, has the time lag been built into the time frame for the project?

Technical support in the form of secretarial or technical assistance:
If these support services are shared, the level of priority for the project and/or the number of hours of secretarial or technical help should be specified.

Political power in the form of the authority to require other personnel to participate in the task as the delegator specifies:
The delegate may form committees, delegate sub-tasks, sign his or her own letters, and represent the manager to outside agencies. It is poor delegation to review and sign letters prepared by the delegate or to brief him or her on what to say at meetings with outside agencies.

Follow up to a Delegated Task:

Good managers use delegation as a way to develop their staff. Feedback on the task is an important responsibility of the manager. This can take place in the following ways:

Public acknowledgement of the work done, e.g., a statement to other people in the organization through a in-house publication or by an announcement at a staff meetings;
Personal thanks for having done the work;
Praise for aspects of the work that were particularly well done;
Constructive feedback to improve certain aspects in the next similar task; and,
A note in the employee's personnel file.

If you take care of the people, the people will take care of you.

Motivation

There are five basic theories of motivation of direct interest to the project manager. Each of them will manifest itself at different times during the project's execution, but the manager can utilize them to better understand why their people are making the choices they are making.

A need is a psychological or physiological deficiency that makes a certain outcome appear more attractive.

A motivation is a willingness to do something conditioned by the outcome of doing that thing.

The basic motivational theories:

Goal Setting Theory

People who set goals are more highly motivated and achieve more.

Reinforcement Theory

Events immediately following a behavior reinforce that behavior.

Attribution Theory

How we judge people based upon how we attribute their behavior:

Distinctiveness

Different behaviors in different situations

Consensus

Everyone behaves similarly in a similar situation

Consistency

In a person's actions

Equity Theory

Employees compare the transaction with similar transactions in the organization.

Expectancy Theory

Most supported by the research

Behavior is shaped by the expectancy of the outcome

Attractiveness

The importance placed on the reward

Performance/Reward link

How closely they are related

Effort-Performance link

Likelihood that extra effort will result in better performance

The Role of Complexity and Chaos in Project Management

Discussions with individuals involved in the Project Management function will often produce the same list of Project Management problems. These problems include items such as the following:

- Creeping Scope (Performance)
- Unrealistic Expectations
- Lack of Planning
- Lack of Control
- Poor Tracking
- Anomalies occurring during the Project
- Chaos throughout the Project
- Others

Given the opportunity, the list could be extended to several pages. What is interesting, however, is the common thread in all the problems listed; that is, the appearance of Chaos within the structure of the project. This Chaos is frequently credited with the translation of a project with an exquisite project plan into a jumbled, virtually unmanageable mess.

Unfortunately, it is not possible to tell you that your projects will be entirely manageable. In fact, what is closer to the truth is the assertion that all projects exist within a chaotic environment, and will experience chaotic periods. That's the bad news. The good news is that Chaos is understandable and manageable on a system level. You can manage a project in a chaotic environment if you are willing to allow the project to act in a natural manner.

What is Chaos and Complexity?

Chaos is a relatively new branch of mathematics that has received a great deal of attention in recent years. One of the more straightforward explanations of the phenomena may be found in Bak (1996). In this profoundly named text, Bak describes from an empirical perspective how the realm of Chaos exists all around us. Bak notes that chaotic models can be used to describe earthquake patterns, the use of English words in this textbook, and the pattern of behavior of the U.S. stock market.

Some have taken issue with Bak's work, stating that his model is nothing more than an empirical description of complex behavior, and is therefore of little use to the practitioner. One need only consider the fact that the majority of our natural sciences knowledge was developed by similar means to recognize the potential fallacy of this argument.

Bak proposes a model of Chaos that exists in the physical domain between order and randomness. This is shown in **Figure 69**.

Order Chaos

Random

Figure 69: Chaos in the Physical World

Obviously there is some hope, as Chaos is not a random process. It is clear that Chaos is more difficult to understand than an ordered model. In fact, Chaos is a phenomenon that exhibits the behavior of a complex system. It is so complex that we are unsure of all the constructs contributing to the behavior.

This type of model reflects the real world of Project Management. In short, Project Management is impacted by a wide variety of sources within an organization, many of which are beyond our reach or awareness. Thus, the study of Chaos will enable us to better understand the overall system of Project Management.

A Simple Chaotic Model: The Sandpile Model

It is beyond the scope of this text to describe in detail how Chaos and complexity function. It is sufficient for the understanding of the project manager to investigate the behavior of two simple Chaotic models. The first of these models is the Sandpile model. The second model is the Metered Traffic model.

The Sandpile is described in detail by Bak (1999, *How Nature Works*. Springer. New York) and can be implemented with a simple deck of cards. An exercise of this type will enable the project manager to very quickly grasp the behavior of a chaotic system. First, we must define some of the behaviors of a complex system:

It operates under a set of simple rules
It self-organizes into a state of criticality
Systems in a state of high criticality exhibit catastrophes

To model a simple sand pile, place nine playing cards on a table in a three by three matrix as shown in **Figure 70**.

Figure 70: Sandpile Matrix

The 3X3 matrix will represent a Sandpile, and the addition of single cards to the system will represent a single grain of sand being dropped onto the Sandpile. As might be expected, when enough grains of sand are dropped into the matrix, some avalanching will occur. This effect is described with a simple set of operational rules for the simulation. These rules are as follows:

1. Introduce a card randomly to the matrix.
2. If there are four cards in any pile, these cards avalanche into the N, S, E, and W piles adjacent to the original pile. This is shown for the center pile in **Figure 71**.
3. If a pile is on the edge of the matrix, cards moving out of the matrix are considered to have exited the system.
4. Record how many cards moved during the process.
5. Continue introducing cards into the matrix until you have exhausted the original deck.

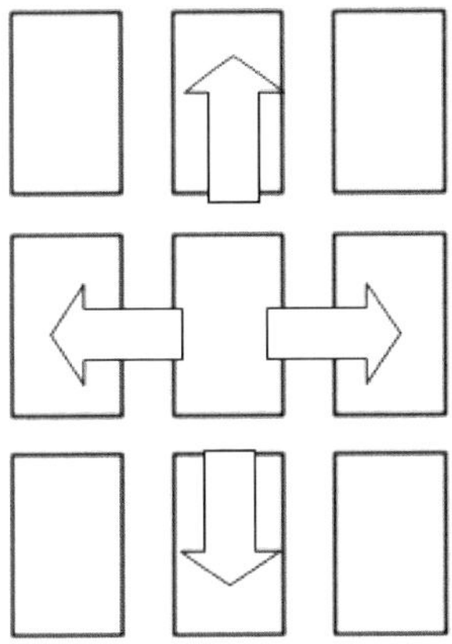

Figure 71: Avalanche from the Center Pile

It becomes very clear to the individual experiencing this simulation that a large number of cards can move with the introduction of a single card. This typically happens only after a number of cards have been introduced into the matrix, thereby increasing the criticality of the system. As criticality of the system reaches a critical mass, avalanches become the norm. These catastrophes or anomalies are therefore present in all systems, and the frequency of their occurrence is dependent upon the critical mass of the system. Remember those anomalies that occurred on your last project? They are a part of every project you will ever manage, and we must discover how to minimize the frequency of occurrence of those anomalies.

The 3X3 Sandpile model is very small, and therefore easily implemented manually. It can be thought to represent the behavior of nine simultaneous projects operating within a single organization. Upon investigation, we discover that the edge effect experienced by the system is large, caused by the fact that eight of the nine cards form an edge of the matrix. Therefore, a larger matrix might be more informative. However, a larger system rapidly becomes unwieldy for the user, and is best modeled on a computer.

It is clear that many grains of sand must be introduced into the system before any avalanches occur. As the critical mass increases, the avalanches become more frequent, until eventually they become the norm. (Note that the numbers recorded in **Figure 72** represent the number of avalanches, representing the movement of four grains of sand.)

What is of interest to the project manager is a perspective on the patterns occurring in a chaotic system such as this. The frequency distribution of a 25X25 matrix is compiled in **Figure 72**.

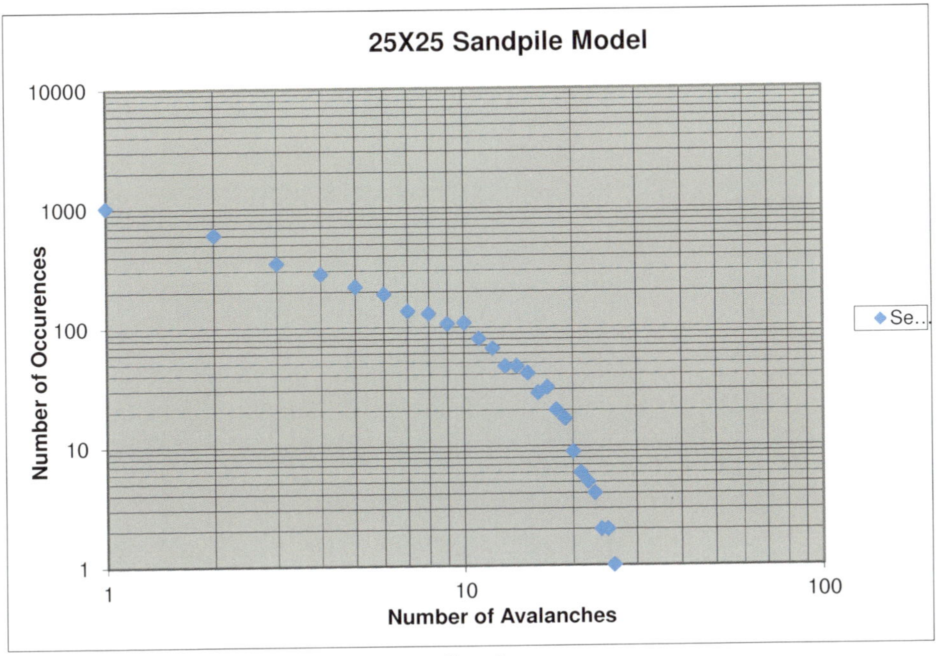

Figure 72: Frequency Distribution

Thus, the behavior of a chaotic system follows a power law. Note that the curving downward of the line is caused by both edge effect and sample size. Larger data sets do not exhibit this phenomenon.

The project exists within a chaotic system. There are numerous other projects within the system, all bumping into each other. The single project system will exhibit many small magnitude catastrophes throughout its duration, as well as fewer large catastrophes. These are the anomalies discussed earlier. They will exist in every project, and we must manage the project to account for their existence.

A Model of Chaos: Metered Traffic

A second model of Chaos is demonstrated in the behavior of metered traffic. What is noteworthy in this example is that a chaotic problem is presented with a solution for handling this problem. The method of solution is applicable for the project manager, as well.

Metropolitan areas frequently experience traffic problems due to excessive volume. One method utilized to alleviate this problem is metering the flow of traffic into the system, as on entrance ramps to freeway systems. Some have argued that this simply transfers the problem from freeways to surface streets. However, as we shall see, this is not the case.

The traffic system operates under simple set of rules. The traffic flow is analyzed, the difference between the actual traffic flow and the maximum possible flow is calculated, and the rate at which cars are metered into the system is set. The design of the system is focused on the maximization of traffic flow through the system. Obviously, this system has all the elements necessary for chaotic behavior.

The system was designed to maximize the flow of traffic, which is essentially maximizing the critical mass of the system. This would imply that the system would be more susceptible to catastrophe brought about by a small disturbance. Remember that the small disturbance in the Sandpile model was the introduction of a single grain of sand to the system. In the traffic case, the disturbances may be sourced from those elements already present in the system, as well as those entering the system. As a tangible representation of an avalanche, imagine what occurs when a driver of one of those cars applies the brakes. The drivers immediately following also apply the brakes, and the disturbance avalanches throughout the system. In this case, a small tap on the brakes might impact ten thousand cars.

The solution presented by the traffic managers was to reduce the critical mass of the system. Instead of metering cars at a maximum capacity level, they chose to meter cars at a seventy-five percent level. This resulted in a buffer zone around each car, so that when brakes were applied it affected far fewer vehicles, perhaps ten cars. Thus, the reduction in criticality of the system resulted in smaller, more manageable catastrophes. There is a counterintuitive conclusion to be drawn from this. *If cars are metered at a less than maximum capacity, more cars get through.* Although waiting time to enter the freeway is increased, overall trip time is reduced in the general case. Consistency is increased, and therefore the quality of the system is increased.

This conclusion was extended to the manufacturing factory floor, where Industrial Engineers reduced the rate of the bottleneck machine in the plant from maximum capacity down to seventy-five percent capacity. Again the result was counterintuitive. They were able to make more parts. *By reducing capacity, they were able to make more parts.* Chaos plays a very significant role in our lives. Lack of planning for chaotic behavior can be a very costly oversight.

As a last note, one can discover many chaotic models about us in the world. These models typically exhibit behavior more complex than is easily understood on a detailed level. The system level is more easily understood, such as in the case of the Windows operating system. The behavior is chaotic, but yields an appropriate complex behavior for the system.

To better understand this model, and to understand the impact it has on project managers, we will examine the results from a simple Sandpile model. In the simplest form of a three by three model, we understand that there are a maximum number of cards that can be placed upon the table at any given time; twenty-seven cards. While this value is possible, it is not extremely probable due to the highly interactive nature of the system. In Figure 73, we have graphed the results of a randomized sample, with the X-axis representing the sequential number of the card added to the system, and the Y-axis referring to the total number of cards on the table. This result and then illustrates for us that the system tends to self-organize into a critical state at approximately fifteen cards on the table. The probability of an avalanche becomes quite high, resulting in cards being ejected from the system and leaving the table.

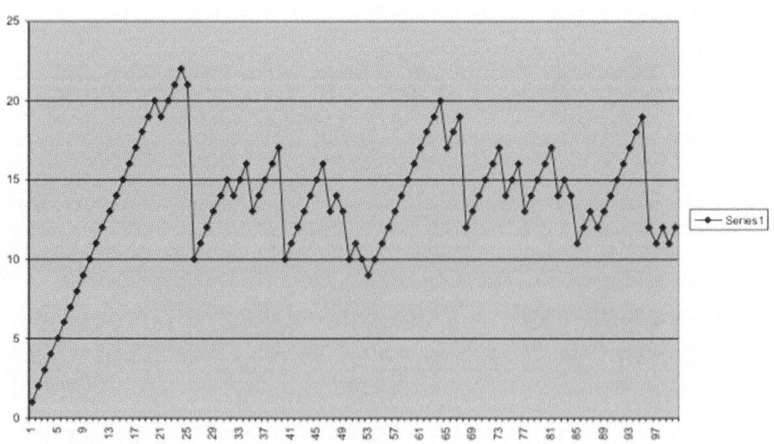

Figure 73: 3X3 Sandpile - Number of Cards on the Table

This result is somewhat interesting, but it does not dramatically illustrate the potential problems this creates for project managers. We know from our consideration of chaotic systems that they will self-organize into a state of critical mass. This means that the system is in a precarious position, and is susceptible to any small disturbance. These small disturbances may result in significant avalanches that will cascade throughout the entire system. These avalanches will typically result in impacts to the project manager's ability to complete the project on specification, schedule and budget.

When we examine a larger system, that of a twenty-five by twenty-five matrix, we find that the system illustrates the same chaotic behaviors but does not exhibit a significant edge effect. This edge effect is what tends to defuse an explosive system in process when an avalanche occurs in a small system. In a larger system, there are many more interacting cells within the matrix, and fewer edge cells. In the case of the three by three matrix, eight of the nine cells are edge cells. In fact, four of the cells are double-edge cells. This means that eighty-eight percent of the cells will cause cards to leave the system within an avalanche occurrence.

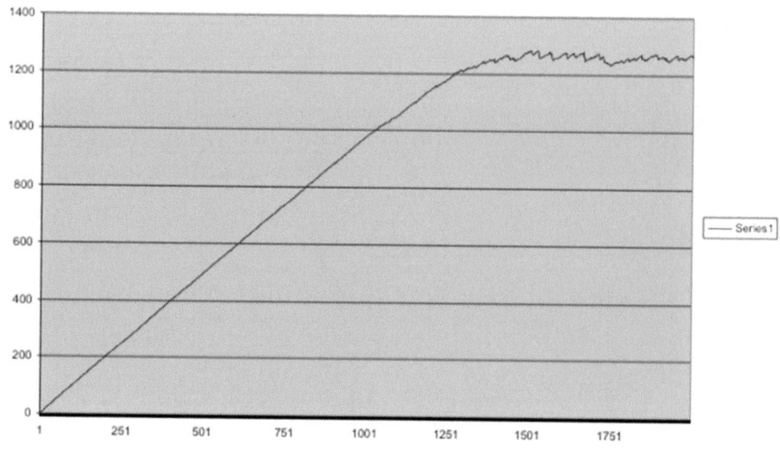

Figure 74: 25X25 Sandpile - Number of Cards on the Table

The twenty-five by twenty-five matrix contains a total of potentially one thousand eight hundred seventy-five cards on the table at any given time. While this value is possible, it again is not probable. As can be seen in Figure 74, this chaotic system will tend to reach an equilibrium point at approximately one thousand two hundred fifty cards on the table. It should be noted on the graph that for a long period of time very little disturbance occurs within the system. However, when the system reaches a point of critical mass, avalanches become commonplace and the system tends to remain at a point of critical mass.

When we consider the impact of the edge effect on this larger system, we recognize that of the six hundred twenty-five cells in the system, only ninety-six contained edges. This means that in the larger matrix, there is approximately a five percent chance of a card's being ejected from the system. Thus, the larger system will demonstrate a closer adherence to critical mass once the system is in operation.

This is a critical insight for the project manager, as it establishes the fact that any chaotic system with significant magnitude will likely remain in a highly precarious state throughout the execution of the projects. Dust Chaos becomes a way of life, and must be dealt with as a necessary consequence of existence within the organization. In order for us to understand how we will operate our projects effectively within the chaotic regime, we must understand with some certainty what expectations can be defined for operation.

We must consider that we are principally interested in the average amount of effort that will be consumed to accomplish a specific amount of work within the organization. If we equate this to the Sandpile model, we can consider that a three by three matrix experiment represents a department made up of nine individuals. As it turns out, each of these individuals is quite effective at managing three tasks simultaneously, but in the event a fourth task is assigned, must find a way to hand off some work in order to maintain appropriate levels of productivity. This creates a definition for our model, a productivity model utilizing the Sandpile.

In this model, we will assume that regardless of the size of the system investigated, each of the individuals within the system will be represented by a single position in the matrix, and that each is capable of accomplishing three tasks simultaneously. If a fourth task is assigned, the individual must prepare his or her work for handoff and suffer a decrease in productivity.

Thus, our productivity model will assess the total number of cards on the table, and reduce this total by a factor representing the loss of productivity; that is, the total number of cards that moved. To examine this productivity model, we ran a randomized Sandpile simulation with a total input of two million cards. The results are summarized in Figure 75.

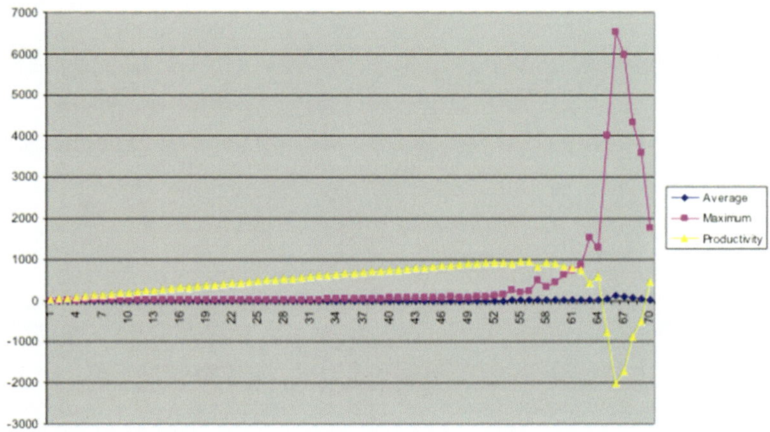

Figure 75: Number of Cards moving - Productivity of the Sandpile

This particular model demonstrates our data in a somewhat modified fashion. In this particular case, the X-axis is represented by the normalized value of the percentage of cards on the table. As an example, if we had one thousand eight hundred seventy-five cards on the surface of the table, this would be represented by a one hundred percent value. We note from the data that we do not exceed seventy-one percent of the potential cards on the table, which corresponds to our equilibrium point of less than one thousand three hundred cards on the table.

The blue line in this diagram represents the average number of cards that have moved. The pink line represents the maximum number of cards that have moved. It should be noted, before leaving this subject, that these values tend to drop off as we approach the right hand side of the data set. This is largely due to the fact that a very low probability of occurrence exists in this particular region due to the precarious nature of the system. This implies that the pink line will continue upward to infinity as we approach the practical maximum for the system. In essence, this illustrates the fact that the closer the system gets to fully loaded, the more cards will move during a typical avalanche event.

The yellow line is defined as total productivity for the system. It will be noted that this line increases steadily until it reaches approximately seventy-five percent of the practical maximum for the system, at which point it begins to decrease, ultimately becoming negative. In practical terms, this means that when we overload our workforce, they tend to accomplish less as they hand off the work to others.

Therefore, it is clear that we have a loss of potential productivity when we are not utilizing seventy-five percent of the total resources available. Likewise, we are losing productivity when we exceed seventy-five percent of the resources readily available. Therefore, to maximize our productivity we should attempt to target a seventy-five percent utilization of our system. Anything less surrenders potential resources that could be applied to the work to be accomplished, and anything more introduces chaotic losses that will consume more resources than the benefit achieved by additional work.

By examining the impact of chaos on the workforce, we have mathematically established that the introduction of disturbances into our system can result in losses in productivity. In addition, overloading the workforce results in further losses to productivity. It is critical for the project manager to recognize these impacts to the system, then, and to plan the work to be accomplished accordingly.

What about the project manager?

Dr. Fred Zimmerman (1997) most succinctly stated the role of the project manager when he noted that the project manager must embrace the fact that as a project manager, he or she will always be in trouble with someone; the question is just with whom. Likewise, Chaos exists on every project. Acceptance of these facts enables the project manager to accept the chaotic behavior of the system as a necessary consequence of operating in an imperfect environment.

If we could insulate every project so that it would never be impacted by other projects within the organization, we could significantly reduce or eliminate Chaos from our list of concerns. However, as Peters (1987) has pointed out, all white-collar work is essentially project-oriented. This implies that all project managers are faced with the likely occurrence of Chaos within their daily activities.

What can be done to manage a project within a chaotic environment? The project manager must recognize that excess resources must be committed to the project in terms of management reserve to account for the complex behavior of the organization. As stated in the traffic example, it was discovered that a committed reduction to seventy-five percent of the resources resulted in superior traffic flow. Thus, the project manager must schedule the resources on the project such that they do not exceed seventy-five percent utilization.

In project terms, this means that three days of effort might be required to complete a specific task. The duration utilized in the budget will accordingly reflect a three-day effort to accomplish the task. The schedule, however, will reflect a four-day completion cycle. This communicates that three days of work will be accomplished, but it will take four days to finish the work. Why is this the case?

Phone calls
Email
Voice mail
Conversations
Interruptions
Other projects
Other people
Anomalies
Others

The counterintuitive conclusion to be drawn from this discussion is:

If you schedule projects for seventy-five percent capacity, you will get more work accomplished.

Project managers will increase their success rates by reducing the rates at which they schedule projects. In many cases, organizations schedule personnel on projects at one hundred percent utilization, or worse yet, one hundred twenty percent utilization. This is accomplished through a company culture that demands mandatory overtime of its employees. Project managers must, therefore, plan for chaotic behavior on their projects by reducing the utilization of the resources to a more realistic level. Is seventy-five percent an appropriate level to compensate for these effects? That is for the project manager to determine in his or her specific organization and application.

It is clear that Chaos can be managed to account for all that occurs during a project that cannot be predicted or incorporated into its project plan. It is time for project managers to become forward thinking, and understand that the anomalies that caused problems on their last projects are ever-present in the system and will appear on their future projects.

Phase 5: Project Close-out

Project close-out is an essential component of any effective project manager's repertoire. When we first undertake to manage projects, we initially create our project plans based upon our best assessments of the durations and efforts necessary to complete the work. If we examine our ability to accurately predict the future based upon this limited experience, we discover that our accuracies fall well below our expectations for successful projects.

If we undertake to improve our performance on projects on a continuous basis, we must expand our knowledge with regard to appropriate techniques and methodologies to complete technical work, and we must develop our self-awareness with regard to our estimating capabilities. Therefore, we utilize the close-out phase of the project to examine our original project plan, and compare that to the actual outcomes we observed on the project. This enables us to improve our ability to estimate and more accurately predict the outcome of the project, if we are honest with ourselves and truly seek to improve our abilities in this area.

The Project Post-Mortem

With all the challenges that project managers face in the development of complex projects, it isn't particularly surprising that very few of us take time to conduct project post-mortems. The function of a post-mortem is to summarize a given project history and analyze all the positive and negative impacts on the project in hopes of developing a predictive view of how our projects will function in the future. Essentially, a project manager is assessing his or her own ability to predict the future and

then perform to that prediction. This assessment will often result in new guidelines or recommendations with regard to Project Management approaches going forward.
An effective post-mortem is typically conducted within several weeks of completion of the project. This allows the individuals conducting and participating in the post-mortem to have fresh memories of the events of the project, and for sufficient separation from the project closure to minimize emotional input.

Here are ten suggestions for conducting a good project postmortem:

1. Involve all the contributors. If others were involved in your project, arrange a project review. If you were the sole contributor, set aside time to reflect on your experience. Document the good, the bad, and the ugly aspects of the project. Note all grievances, and solicit ideas for future improvement.

2. Document the post-mortem in writing. The document may then be published so it is easily accessible to everyone in the company. A series of post-mortems can ultimately evolve into a valuable record of your expected performance.

3. Begin with a project overview. Begin the post-mortem document with a brief overview of your project, including a design overview, estimated budget, and project start and finish dates. Describe the product's purpose, intended customer, and other general information.

4. Include project details. Document the quantifiable details of your project. Include your schedule and budget baseline, as well as the true outcomes. List everyone who worked on the project, and document his or her contributions.

5. Document what went right. Did the final product turn out better than expected? Did you experience an occasional stroke of luck? List at least ten things that turned out well for your project.

6. Document what went wrong. What difficulties did you encounter? What assumptions proved to be incorrect? Compile a detailed list of your project's shortcomings. List at least ten things that turned out worse than expected. Document unpleasant mid-project surprises, as well.

7. Assess your risk management. Document all the risks you took during the project, and note how effectively you managed them. Did you explore experimental ideas, use new technology or development tools, or develop for a new platform? Did your schedule slip from poor estimation practices? Did you take big risks, or did you play it relatively safe? Assess your risk management experience, paying particular attention to how you would change your approach for the next project.

8. Assess your cost and schedule performance. How did you perform on schedule variance and cost variance? Using earned value analysis may provide excellent predictive statistics for use on your next project. The same statistics can often be used as corrective factors on other people's projects, as well.

9. Draw meaningful conclusions. What conclusions can you draw from your project history that will help you improve in the future? What should you start doing, keep doing, or stop doing? Describe new practices you should try for your next project. This is perhaps the most important part of the post-mortem, so be prepared to spend more time here than in any other section.

10. Take action. The final step is to develop an action plan that may be applied to your next project. Review your project history, reflect on the lessons you've learned, and specify new guidelines to follow in the future. Create a checklist for your next project. If you already have a checklist, update it with new refinements. A final benefit of conducting a post-mortem is that you gain a very gratifying sense of closure through documenting all the personal growth you've experienced as a result of completing the project.

So what should be on the list of questions? I've provided some of my favorites below.

General Questions

1. Are you satisfied with the results of the project?
2. What was most frustrating about the project?
3. What was the best part of the project?
4. How should we do things differently next time?
5. What methods do you like the best in our project development?
6. What methods do you like the least in our project development?
7. What three things would you change on this project if you were to start over today?
8. How did our sponsors do?

Project Initiation

1. Did we understand the strategy of the project?
2. Did we understand our alignment with the corporate strategy?
3. Did we identify all the general deliverables on the project effectively?
4. Did we make assumptions on deliverables that we should not have made?
5. Did we get the team engaged before we begin the planning process?

Project Planning

1. How accurate was our original estimate?
2. How could we have improved our estimation process?
3. Does our original specification describe what we ultimately accomplished?
4. What could we do (or not do) that would improve our specification in the future?
5. Did we complete a network diagram describing the inter relationships of the functions of the project?
6. Did our work breakdown structure describe what was accomplished on the project?
7. Did we have the right people assigned to the project roles?
8. How can we make sure we get the right people assigned next time?
9. Did we recognize any problems during the planning phase that occurred later in the project?

10. Could we have completed this project in less time for less cost given what we know today? How can we incorporate these lessons into our future project?
11. Were all of the assumptions made clear to everyone on the project?
12. Did we go down some wrong paths?
13. Did we negotiate effectively during the project-planning phase?
14. Was there anyone missing from the project kickoff who should have been involved from its outset?
15. Were the roles and responsibilities for all of the team members and stakeholders clearly defined and communicated?
16. Were the specifications, milestones, schedule, budget, user needs, and deliverables all clearly communicated?

Project Execution

1. What was the final quality of our deliverables? Could we improve on this?
2. Did the project culminate with innovative and creative solutions?
3. Did everyone have input into the project outcome? Did we engage all the players?
4. Did we have effective project and reviews?
5. Did we receive critical inputs from outside parties positively? What can we do to improve this?
6. Did we treat our team members with courtesy and respect?
7. Does the team want to work on your next project? Why? Why not?

Project Control

1. Did we do a good job overall?
2. What could we have improved on this project?
3. Were we able to ascertain the status of the project at any given time?
4. Were our project predictions accurate?
5. Were we able to do what we said we would do?
6. How can we take the performance on this project to predict performance on our next project?
7. What methods or processes helped you better understand the status of the project and which did not?
8. If our project deviated from the specified path, were we able to bring it back in line with expectations?
9. Did we manage expectations effectively?
10. Did we succeed? Why? Why not?

Project Close-out

1. Did we have the right members on the team?
2. Did we have the right customers in user needs?
3. How could we do better in the future?
4. Did we understand our customer?
5. Were our assumptions representative of our customer's needs?
6. Did our project ultimately meet our customer's needs?
7. Did we miss the mark in any way?
8. Were we able to manage our project effectively during the execution and control phases?
9. Did we get timely feedback on design reviews and oversight from our sponsors?

10. How did the sponsors do overall?
11. How did the team perform?
12. What have we learned about the team and what would we do differently?
13. Was our original strategy effective?
14. Is the customer coming back? Why or why not?
15. How will the lessons learned on this project be communicated to the next?

The Project Management Knowledge Areas

The project management book of knowledge has been generated by a large field of experts in the development of a standardized approach to project management, and is constructed in nine major knowledge areas. The knowledge areas were required to ensure that the document covers all practical domains required of a project manager, providing information and guidance with regard to the completion of his or her projects in a successful manner. The nine knowledge areas are:

Integration Management
Scope Management
Time Management
Cost Management
Quality Management
Human resources Management
Communication Management
Risk Management
Procurement Management

We will deal with each of these in turn to clarify the areas of interest in each knowledge area. We will also find the considerations that the project manager must keep in mind in each. Completing this, we will expand this model to include additional areas of interest, hopefully justifying them regarding the importance of the human factors associated with project execution.

Integration Management

Integration management concerns itself with the coordination of all the project management knowledge areas throughout the project life cycle. If we accomplish integration effectively, we ensure that all of the elements of the project are synchronized in order to maximize our probability of project success.

This area of expertise is not to be confused with Program Management, which focuses on the interaction between multiple projects within the organization. Further discussion of Program Management is to be found in the section of that name.

Integration management usually covers a number of standard processes. These include the following:

Develop the project charter

The project charter will typically be undertaken during the project initiation phase. However, the charter will not be completed until the project planning is complete, implying that it cannot be completed until the end of the project planning phase. The project charter is utilized throughout the project life cycle to ensure that proper authorizations have been provided for the project.

Define the project scope

After developing an extensive list of users and user needs, the initial scope for the project can be defined in terms of capabilities to be delivered, services, or other items. While the scope will not be complete during this initial assessment, it is possible for the project manager to define at a high-level what the major system requirements will be for the project.

Develop the project plan

This area includes the coordination required to accomplish all the planning efforts to produce a coherent specification, schedule and budget.

Execute the project

Utilizing the project plan as a guideline, execute all the activities defined by the tasks on the schedule as well as all supporting tasks necessary to complete the project plan as specified.

Control and status the project

Utilize the project plan to produce a status of the current state of work being accomplished and communicate that effectively to all stakeholders on the project. Utilize those metrics as defined in order to predict the final outcome of the project.

Manage change

Incorporate necessary changes to the scope, specification, schedule and budget of the project during the execution phase. The scope changes are largely brought about by changes in understanding on the project, and will typically results in a rebalancing of resources as necessary to complete the work.

Close out the project

Complete all documentation and assessments related to the close-out of the project.

As can be seen by these descriptions, integration management largely focuses on managing the multiple stages of the project life cycle in a manner that will result in the seamless integration of work and final end products. This implies that the project manager coordinates all of the resources, plans, and tasks required to complete the project as planned. In addition, he or she must ensure that the project retains its strategic alignment for the organization and results in satisfactory performance for the customer.

The project manager must be able to recognize where the interfaces occur on the project, then, and manage the interactions between the disparate entities. Unfortunately, the number of interfaces may increase beyond a manageable level as additional stakeholders and users are brought on board with regard to their interests in the project. Therefore, the project manager must rely on his or her established communications plan and maintain positive relationships across all organizational interfaces.

Throughout the project life cycle the project manager will also develop, document and maintain all appropriate management documentation for the project. Thus, he or she will begin the development of the project charter, and will be responsible for possessing and managing the life cycle of the charter throughout the project. This will include retaining all original copies of the project charter documentation, including signed off copies from the various stakeholders who will from time to time be asked to approve subsequent stages of the project. At the completion of the project, the project manager is responsible for the appropriate archiving of all these documents in order to ensure that they can be reconstituted in the event of a future need.

The establishment of a preliminary scope for the project will require a scope statement or scope document to establish a common understanding of the scope of the project at the outset of the completion of work. This document will become increasingly important throughout the life cycle of the project, as scope creep will dramatically impact the original scope. Because of this, we must ensure that appropriate schedule and budget changes are also incorporated. The project manager must also maintain the original scope statement and signatures as originally understood, should they become necessary to justify increases in either schedule or budget in the future.

The scope description or scope statement will at first be used to establish the boundaries of the project in terms of resources and deliverables. This is often useful to check the understanding of the various stakeholders with regard to the intended direction of the project. It is not unusual at this stage to discover that many of the stakeholders have different and contradictory expectations with regard to what will likely be completed during execution of this project.

The project manager is responsible for the development of an appropriate project plan that will describe the specification, schedule and budget. These documents are interrelated and form a single document for the project manager's use. Often they are considered to be three independent documents, only related by means of the commonality of the project. However, when specification is defined, it will define the schedule and budget for the project, given a few assumptions. Therefore, the project manager is responsible to ensure that the overall project plan remains coherent throughout the life cycle of the project. This coherence means that all three elements of the project plan are in concert and result in a realistic and achievable plan.

In addition to the specification, schedule and budget, the project manager will also establish a communications plan and a business rhythm for the project. If necessary, organizational charts may be generated to ensure that all stakeholders understand the responsible parties for each stage of the project. In highly complex projects a responsibility matrix may also be generated in order to establish the parties who are responsible, accountable, supporting, consulted or informed with regard to elements of the project. All of these documents reside with the project manager during the project life cycle, and are typically archived together at the completion of the project.

Integration management becomes extremely difficult during the execution stage of a typical project. This is where the majority of the time and money will be spent, and where the majority of the stakeholders will interact with regard to the progress on the project. By virtue of this, this is also where the Chaos typically expands exponentially, requiring careful attention by the project manager to ensure that all appropriate integration activities are accomplished. This phase of the project will include managing all of the stakeholder relationships in order to ensure that the project plan is executed successfully. This implies that the project manager will address all of the human needs on the project.

First and foremost, the project manager must ensure the health and wellbeing of the development team, ensuring that they are largely allowed to complete their work without undue interference from other members of the organization. This requires a highly effective project communication plan, as well as a rigorous implementation of a business rhythm. In general, when communications are maximized regarding the information dissemination on a project, meddling is kept to a minimum. Meddling in the project will frequently result in frustrations and irritations on the part of the development team.

Executing appropriate project control and managing the metrics on a timely basis will go a long way towards relieving this pressure from the project team. Therefore, the project manager must take responsibility for the establishment and maintenance of appropriate project metrics throughout the life cycle of the development. Communication on a very regular basis with a great deal of repetition will reduce significant involvement by leadership on the project. Interestingly, many project managers feel that they need only provide updates when the project has challenges or issues to overcome. The project sponsors, on the other hand, feel that if communications are few and far between, there might be problems that the project manager is not willing to disclose. Therefore, the project manager must recognize that regular and frequent communications are the best remedy for too much attention from leadership during the life cycle of the project.

As scope changes are requested or dictated, the project manager must manage these changes effectively, in order to ensure that the project team does not feel they are being treated in a disrespectful manner. This requires appropriate communication and political skills on the part of the project manager. Many of these factors are typically rolled into a single descriptor of leadership. In this case, leadership involves the ability to maintain positive relationships with all of the stakeholders and to ensure that the development team is shielded from undue influence from outside of the project development effort. If this is accomplished effectively, it goes a long way towards engaging the project team in building significant trust in project leadership.

Scope Management

Project scope management largely focuses efforts on scope definition and control during the project life cycle.

Scope refers to the entire effort involved in creating the output of the project and the necessary support processes that are required in order to accomplish the tasks defined.

A deliverable is a tangible product that is output from the project, usually hardware or software or process oriented.

There are several elements associated with typical project scope management including:

Scope planning
Scope definition
Creation of the work breakdown structure
Scope verification
Scope control

The scope management plan begins with the development of a project charter, including a preliminary scope statement as a component of the project plan. The project scope statement defines the basis for defining the project's boundaries. The statement will typically include:

A generalized description of the project
Project objectives
A detailed description of deliverables
Any defining characteristics, requirements or constraints associated with the project

A scope statement is useful to document the boundaries of the project, as well as to define the project's success. Included in the project charter will be appropriate constraints and assumptions, and the verification plan will include assessments associated with acceptability of the solution. Highly detailed scope statements may also include definitions of the cost estimation and project control techniques to be used during the execution phase of the project.

The scope planning and scope definition phase are used to begin fleshing out the project charter at the outset of the project. Ultimately, the project charter will become the project scope statement.

The definition of project also includes the work breakdown structure. This is a construct that will develop and communicate a hierarchical list of tasks that must be completed in order to achieve the project's deliverables. This work breakdown structure will ultimately be used to develop the project's schedule by associating the various tasks with one another and adding durations to the execution of the tasks. Several methods for the development of the work breakdown structure and schedule have been previously presented in this document.

On highly complex projects, it may be appropriate to develop a work breakdown structure dictionary, which provides detailed information about the terms utilized in the construction of the work breakdown structure. This is produced to ensure that any nebulous terms are defined appropriately, to create a coherent understanding of the tasks to be accomplished.

The general construction of a work breakdown structure should focus on the deliverables to be produced by the project. A simple method to ascertain the coherence of a work breakdown structure is to begin by listing all of the deliverables necessary for completion of the project. Once these deliverables are in place, they become the exit criteria for each of the subprojects. Appropriate tasks and processes that must be completed in order to produce the project deliverable are then inserted into the project schedule, defining the work to be completed in order to achieve each deliverable. Finally, the various project deliverables are related to one another by means of interrelationships and chronological relationships in order to complete the project schedule.

Scope verification is essentially the process of ascertaining whether the project charter is a coherent document and whether the described work is objective and achievable. This requires a careful assessment of the completed project charter to ensure that the customer will be provided what they require, as well as whether the project team has the appropriate resources allocated in order to have a high probability of producing the desired products and services.

Scope control is typically focused on minimizing the changes to scope throughout the life cycle of the project. This, however, can be misleading in many Project Management textbooks. It is important to understand that due to changes in understanding and insights on the project, scope control is a necessary component of any project manager's repertoire. The idea is not to limit changes to the scope, but rather, to incorporate them in a manner that will continue to provide a high probability of completion of the project on plan. Therefore, the project manager should not seek to minimize scope changes, but rather, to utilize them to better increase the fidelity and resolution of his or her remaining project plan to make more representative predictions with regard to expected outcomes.

In fact, highly proficient project managers recognize that scope changes allow them to revise their current project schedule and budget, often bringing their project closer to plan than prior to the requested scope change. Therefore, scope changes can be used in essence to re-baseline the plan periodically throughout its life cycle.

A significant component of scope management is the ability to manage requirements throughout the project life cycle. This entails the process of developing complete requirements lists, as well as managing volatile requirements as scope changes are incorporated. In general, it is recommended that the project manager develop and follow a requirements management process in order to manage this complex task.

Most importantly, the project manager must always keep in mind that "if it is not written down, it never happened" on his or her project. Therefore, project managers must document everything carefully, and maintain accurate records throughout the life cycles of projects.

Time Management

Project time management is typically associated with developing an accurate project schedule and maintaining its accuracy throughout the project life cycle. This is of significant importance to the majority of project managers, as they often state that delivering projects on time is the most significant challenge they will face. Schedule conflicts often are created on the project as well, and as the project life cycle nears its completion these conflicts will become more commonplace and more severe.

The project time management processes include:

Task definition
Task sequencing
Task estimating
Schedule development
Schedule control

A task is a work element that has carefully defined contents, as well as an associated duration. This duration will change dependent upon the number of its assigned resources, but from the perspective of the project manager the duration should be associated with the entire work effort.

When examining a typical hierarchical work breakdown structure, the tasks are the work packages that are located at the lowest level of indentation on the hierarchy. Therefore, when a project is constructed of multiple subprojects, the project manager must understand that these are only summary tasks, essentially locating the work within the structure of the project, and are not actually tasks to be completed. When the subprojects are divided into individual work assignments, these assignments or work packages are the tasks where resources are consumed and money is spent.

Summary tasks are used to collect together lower-level tasks, and are simply a mathematical summation of the work packages included within them. This is an important distinction, as the project manager must ensure that resources are only applied to the work package tasks; the lowest level tasks of the project. All other tasks are only placeholders and constructs of the hierarchy, and any resource allocation associated with them is superfluous and is typically not recommended.

The task list for a project includes all work that must be completed. However, because its generation is a very complex challenge, different methods may be used to create an appropriate and complete task list. As has been previously discussed, a project manager defines the project's deliverables and identifies the tasks necessary to provide them. This may be accomplished in a top-down fashion, where the major subprojects are defined and then increasing levels of detail are added below those levels, or it may be constructed from the bottom up, where deliverables imply tasking, or the major subprojects.

For this reason, it is common for a hierarchy to be generated describing the tasks to be accomplished in the form of a work breakdown structure. It is not uncommon for the work breakdown structure to contain twice as many elements as there are work packages, due to the need to provide summary tasks and subprojects in order to understand the hierarchy of the project work.

A milestone is defined as a significant event without duration. In general, it is useful for the project manager to consider a milestone to be associated with the delivery of every project deliverable. This, then, becomes the flag that is accomplished at the finish of a deliverable, and creates a significant marker in a typical project plan. For this reason, one should consider that every subproject is completed with a milestone. It occurs immediately subsequent to the project deliverable generated by that subproject. In addition to these milestones, further milestones may be utilized to identify and record the completion of integration activities as the various budget deliverables are incorporated into a completed project.

Once the tasks have been identified, they are interrelated by means of a network diagram or a Gantt chart. These techniques are typically utilized in order to translate the work to be completed into a chronological relationship, allowing the project team members to understand the flow expected in the development of the work on this project. Task sequencing is the process of identifying which tasks must be completed prior to subsequent tasks. This is, of course, a highly volatile area, as when resources are added or reduced on the project they may impact these sequences significantly.

Once the work breakdown structure and test sequencing have been identified, each of the work packages must be estimated for both the type of resource that is required to complete the task and as the expected duration. This might be affected by the availability of resources or the amount of resources available for assignment to the work. The duration specified on the Gantt chart should reflect the actual calendar duration for the completion of that task. Therefore, on a task with two resources, if one resource is removed it is expected that the duration of that task would be increased by one hundred percent.

After completion of the durations and interdependencies on the Gantt chart, it is common to incorporate risk into the project schedule to represent uncertainties in the estimation process. There are several methods that are utilized to accomplish this, including the critical path method and modified versions of the same, in order to provide additional float time within the project. They enable the project manager to utilize his or her discretion in the consumption of the excess time in the form of management reserve.

It is typically recommended that a risk element be incorporated into every major subproject on the project plan, thereby allowing some flexibility for the project manager as differing deliverables are delivered ahead of or behind schedule. This additional float should not be considered as a fudge factor, but rather, as a smoothing factor that is utilized in order to cover minor fluctuations and noise in the development of project deliverables.

Another factor of interest to project managers is the definition of the critical path on the project. This is normally associated with the longest path from the start to the finish of the project, and is much more clearly defined on a typical network diagram than on the Gantt chart. However, most software that is utilized to develop Gantt chart documentation also incorporates simple methods by which the critical path may be defined. This is of great interest, because any delay that occurs in a task on the critical path will result in an equivalent delay in the delivery of the project. Therefore, project managers spend the majority of their time managing the tasks that are on the critical path. As might be expected, when there are fluctuations in the completion of specific tasks and subprojects on the project, elements will enter and exit the critical path as they become the tallest pole in the tent.

Creating realistic and effective schedules is the largest challenge faced by project managers. Human beings have a tendency to view the project scheduling activity with an optimistic eye, resulting in unrealistic schedules and expectations. The project manager should always keep in mind that he or she is paid for accurate prediction of the future, and one of the most important elements of this is ability to produce predictive project schedules. Therefore, it is recommended that project managers pay close attention to the original estimates they have made for given projects, and to comparing the actual results of their projects in order to improve their future estimates. This will be one of the most significant things that a project manager can utilize to improve his or her ability over time with regard to the predictability of project schedules. Always remember that a project team should not be expected to pick up the slack for poor project estimation.

Schedule control is normally associated with assessing the current status of the project and making a prediction with regard to the expected outcome. This is fundamentally an extrapolation of the existing data; therefore the accuracy of the prediction improves as the percentage of completed project increases. Various methods are utilized to accomplish project control, with earned value management systems being one of the most readily accepted in industry.

A final consideration regarding the use of software in project management: most project managers feel that the use of project software is the centerpiece in the completion of his or her project duties. While this software may be very useful in understanding the interrelationships of the testing on the project, as well as in the generation of graphically acceptable schedules, the software does virtually nothing with regard to enabling the project manager to produce realistic schedules or manage the project control effectively during the project life cycle. Therefore, recognize that project software is nothing more than a tool utilized to represent the information generated by the project manager and team, and is only as effective as the effort put into generating predictive information.

Cost Management

Project cost management is the primary responsibility of the project manager during the execution phase of the project. This includes the cost estimating, budgeting, and cost control throughout the project life cycle. Technically, project managers are responsible for the overall cost of the project, which include tracking and managing costs across all five phases. Since the majority of the resources are consumed during the execution phase, this is where the majority of their attention should reside.

During the execution phase of the project, project managers are responsible to ensure that the project will meet both the plan schedule as well as the plan budget while achieving all of the project deliverables as specified in the project plan. This implies that the project manager must be aware of the current status of the project and be able to make accurate predictions of the likely outcome of the project during the project execution phase.

In order to accomplish this, all necessary factors relating to the status of the project are translated into a common factor, typically dollars or euros. This is necessary in order to be able to make comparisons across the specification, schedule and budget, thereby providing appropriate trade-offs based on the impacts of changes in these areas.

To convert specifications into dollars, the project manager must consider what the cost of not achieving a specific requirement might be, or the potential cost of exceeding a given requirement. To convert schedule items into dollars, he or she needs to consider the resource applied to that item, the effective burden rate of that resource, and the duration over which the schedule is being adjusted. This allows all elements of the project to be described and assessed using a common unit.

Direct costs are the costs on the project that can be directly related to producing the deliverables. Indirect costs are those that cannot be directly related to the production of deliverables, but are indirectly related to the effective completion of the project. Examples of indirect costs might include administrative or IT support throughout the duration of the project. Sunk costs are those where the money has already been spent, and is typically not used in the predictive assessment of the project.

Project managers learn very quickly that they must make extremely accurate cost estimates in order to increase the probability of success on meeting their project plan. As they are using these project estimates as elements of their prediction for the outcomes of projects, it is very important that they take them very seriously and spend appropriate time to ensure exactness. Without this, their predictions are likely to be misleading at best, and will typically not result in effective project management.

Cost estimates are similar to other estimates on the project; they are relatively vague and inaccurate at the outset of the project, and as knowledge increases throughout the execution of the project they may be revised to add fidelity and resolution. The project manager should accomplish these revisions on the regular basis, in order to maximize his or her ability to make predictive decisions.

The project Gantt chart includes the major components of the project schedule as direct costs and labor costs. The project manager will utilize this project plan in order to generate an expected cost curve for the overall project. Cost control is usually nothing more than comparing the expected cost curve to the actual cost curve, and utilizing the discrepancies between the two as a predictor of final project outcome. Unfortunately, the effective use of costs as a predictor implies the use of accurate timecards and reporting on the project as described in the earned value management system. This can be a significant problem for many organizations, but it need not challenge the project manager, as many appropriate timecard systems are nothing more than simple spreadsheets utilized throughout the various departments. It is recommended that the project team members report their time on a frequent basis, such as daily and certainly no more than weekly, in order to ensure that the project status is accurately represented.

There are many common problems that occur with regard to project estimates. One of these is the propensity for many project managers to attempt to manage their projects utilizing tasks that are of too long duration. For example, if a project manager allows a task to be generated on his or her project plan with a duration of sixty days, it is often nearly impossible for him or her to ascertain the task status within the sixty-day period. This implies that the project manager can only know the status of his or her project plus or minus one task duration. For this reason, project managers must ensure they have appropriate resolution in their project schedules. In general, no single task should exceed forty hours on a well-managed project.

At the outset of the execution phase, the project manager will establish a project baseline. This will include a schedule baseline as well as a budget baseline. Cost control will be based upon these two baselines, enabling the project manager to ascertain the schedule performance as well as cost performance at any time during the execution of the project.

Quality Management

Project quality management is another responsibility of the project manager. In general, quality is defined as a conformance to requirements. Upon consideration, it is apparent that the project manager is responsible for the project's maintaining coherence with all of its requirements. Therefore, he or she must be focused on adhering to requirements in order to create a compliant project at the completion of the development effort. A compliant project typically refers to one that meets all of the requirements specified in the project plan.

Quality can also be used as a term to describe fitness for use. This is a much more nebulous concept, as it is very difficult to ascertain the validity of the project approach until after completion of the initial articles to be delivered. This does imply that there are two major areas of interest that must be completed in order to provide an acceptable outcome. Thus, the project manager must ensure that requirements are met as well as guaranteeing that the needs of the customer are addressed satisfactorily.

The project quality system typically includes quality planning and control. Quality planning includes all attempts by the project manager to ensure that the requirements are appropriately documented and that completion of the requirements is verified by means of a project verification plan. The generation of the project verification plan will be the most significant outcome of the quality planning process.

Once the project has defined all appropriate requirements for completion of the development effort, the project team is responsible for devising how each of those requirements will be verified at the completion of the project. Typically these verification methods will fall into four different approaches:

Inspection
Analysis
Testing
Demonstration

Therefore, if the requirement is defined for the project, which might include a performance characteristic, the verification plan should define how that performance characteristic would be ascertained and proven.

As an example, imagine that we are developing a new coffee maker for use in the home. We may set a requirement which suggests that the coffee maker should draw no more than 4 A of power from a standard 120 V AC circuit. To verify this requirement, we may provide an analysis of the power system designed within the device. We may also choose to do a test, measuring the actual power consumption during the operation of the system. Either or both of these methods would be an acceptable verification to complete the project verification plan.

From the perspective of the Project Management Institute, the quality on the project extends well beyond the meeting of requirements and the verification of the same. The quality planning for the project in this domain is in reference to managing the overall enterprise quality for the organization. This might include assessing how this project fits into the overall quality management system for the organization, and establishing what impacts the project may have on the organization from that perspective.

When considered in this broader context, we see that project quality can refer to how the project will operate within an existing Six Sigma process in an organization or in the midst of a lean manufacturing environment.

For example, early in the development of a new technology, it is common practice to allow the development team to violate some quality principles in order to allow the completion of the development work. If this is the case, it is also expected that the project team will clarify when the project will begin to adhere to the appropriate quality management system within the organization, ensuring that appropriate company standards are accomplished by the end of the development phase.

Six Sigma is a very common quality management approach in use in many of our organizations. It is defined as a comprehensive and flexible system for achieving, sustaining, and maximizing business success. Six Sigma is driven by careful understanding of the customer's needs, disciplined use of facts, data, and statistical analysis, and diligent attention to managing, improving, and reinventing business processes.

The basic premise of the Six Sigma approach is that it is not adequate for us to achieve a high level of quality; the expectations of our customer are typically much higher than what we consider to be necessary. For example, we might consider that a system is operating acceptably if it is up more than ninety-nine percent of the time. If we consider this performance from the perspective of the user, though, we might discover a higher expectation. In general, Six Sigma targets 3.4 defects per million occurrences as a basic acceptance level for the performance of any system.

If any shortcomings exist on the project and quality improvement is required, it is recommended that the project manager utilize the DMAIC process for quality improvement. This process includes:

Define: define in detail the problem and opportunity, process, and all associated requirements for successful implementation of the system

Measure: establish appropriate measures for the system, collect measures, and analyze the data in a manner that generates appropriate information about the system

Analyze: examine the information provided by the measurement step to identify and quantify potential improvement opportunities

Improve: generate technical approaches for improvements with regard to the problem or opportunity

Control: track and verify the validity of the improvements implemented

Six Sigma quality systems have been used with great success in many large organizations around the world. Their use has become more widespread as the benefits of an effective quality system have been disseminated, and most organizations take advantage this approach. An effective project manager must be cognizant of these systems in order to plan and implement his or her projects in concert with the systems.

Many Quality texts exist, and the details of these systems extend well beyond the scope of this text. It will be left to the reader to identify and peruse these texts in order to enhance their knowledge of Quality systems.

Human Resources Management

Project human resources management focuses on how resources are planned and implemented on the project during the execution of the development phase. It includes areas of resource estimation, getting people assigned to the team, and managing the team during the execution phase.

Areas of specific interest to the project manager include:

Motivation theories
Influence managements
Productivity

There are many studies related to the motivation of humans, focusing on a variety of methodologies for producing effective behavior in the personnel we are managing.

The first of these is Maslow's hierarchy of needs. In this theory, Maslow postulates that individuals have a variety of needs that must be fulfilled in a stepwise fashion to achieve higher levels of motivation.
The hierarchy is seen in **Figure 76**.

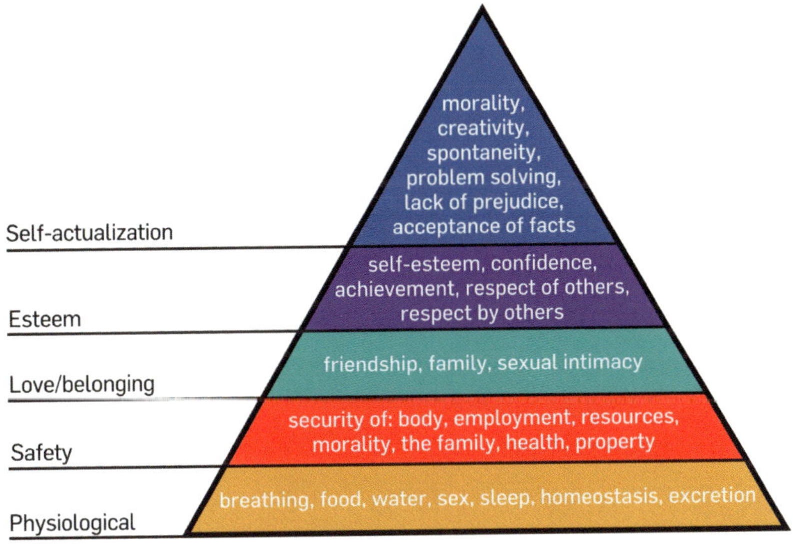

Figure 76: Maslow's Hierarchy of Needs

Self-actualization – when individuals reach this highest level of motivation, they are achieving all that they can achieve. In general, this reflects individuals who are highly engaged in and entirely passionate about their work.

Self-esteem – individuals who are motivated by their self-esteem needs often seek recognition at work, titles at work and any other physical manifestation of their stature within the organization. This might include an interest in offices and authority over other personnel.

Social – individuals who are motivated at this level are often interested in the social aspects of the workplace. This might include the friendships that they have formed with other colleagues in the workplace and the familial atmosphere that is generated in many organizations.

Safety – individuals who are motivated by their safety needs are often under duress within the organization. These are individuals who feel threatened within the workplace, either by their leaders actions or by events taking place within the organization, such as layoffs.

Physiological – the basic human needs that must be fulfilled for any individual to be even remotely interested in his or her work. They include things like air, food, water, interaction, and rewards within the organization.

Herzberg's motivational theory focuses around two fundamental factors related to motivation. The first is hygiene factors; or large salaries, supervision opportunities, attractive work environments, celebrations, food and accouterment provided by the organization. In general, these do not provide increase productivity within the workplace. However, the lack of these items can cause significant dissatisfaction, and will thereby detract from performance. Motivational factors include achievement, recognition, rewarding work, responsibility, advancements, and growth. All of these are directly related to an increase in job satisfaction.

Douglas McGregor theorized that there are several management methods of interest to any project manager, including:

Theory X – workers are generally bad, and they dislike and avoid work so managers must constantly oversee them. This is where the term supervision comes from.

Theory Y – workers are generally good, and the employees consider work as a necessary component of success and self-fulfillment. In general, these individuals do not need a great deal of supervision.

Theory Z – workers not only have a desire to do well; they have a need to do well in the workplace. By virtue of this, they require only the tools necessary to accomplish their work to produce exceptional results.

In general, project managers will be more successful if they are effective listeners, and are willing to build significant relationships with the individuals operating on their projects. This requires a significant investment of time, but also results in greater flexibility during the execution phase of the project. Employees who have a strong relationship with the project manager are more likely to provide the benefit of the doubt to the project manager when under duress, and will often increase their productivity levels if necessary to support a project manager with whom they feel engaged.

Project managers must also engage in organizational planning. This includes identifying and documenting the various project roles and responsibilities to ensure that everyone is aware of the work they are expected to complete in the accountability requirements of that work. This organization might include common structures such as a team organization, a self-directed work team, and integrated product team (IPT), or other constructs that lead to effectiveness in the development work. It is largely at the project leader's discretion to define the organizational constructs that will be most beneficial to complete the work.

Some project managers find it useful to create a responsibility assignments matrix to delineate the various work assignments and accountability requirements for the entire team. This may be useful in cases where conflict exists over responsibilities or when individuals seek recognition for their work. There is some value in using an RASCI chart if this issue persists. This chart includes the assignment of:

Responsibility
Accountability
Supports
Consultation
Informed

For any given task a single individual is assigned to accountability and responsibility. Various other individuals may be assigned consultation or may simply be informed with regard to that particular task. This is a useful manner to clarify the overall responsibilities of different individuals or groups on the project plan.

The project manager is also responsible for developing and

maintaining the staffing plan for the project. This is typically extracted from the Gantt chart, where the information is clarified by showing what resources are required when on the project. This type of resource chart may be very useful in identifying over- and under-allocations of personnel, and will enable the project manager to resource level those challenges in an effort to increase the probability of success.

Once the project manager has developed the completed project plan, which includes a Gantt chart with resource allocations, he or she can work together with the functional management groups within the organization to get individual personnel assigned to the project. This will result in a great deal of conflict and conversation between the project managers and the functional managers. It also defines the allocation of personnel within the organization, and based upon their availability, defines the areas of float required to keep the project on track. Therefore, although the project manager may have created an elegant schedule that shows a completion date on December 1, the inability to dictate when resources are available may result in the schedule being extended to delivery sometime in February. This is essentially a resource-leveling step, and it is crucial for all within the organization to understand the allocation of resources against the needs of the project.

In addition to the resource planning on the project, the project manager is also responsible for resource control and development during the execution phase. Resource development often means identifying team members' shortcomings or challenges, and providing opportunities for appropriate skills development in the form of training or other educational opportunities. This can be extremely time consuming; nonetheless, it generates a great deal of worker engagement if it is undertaken effectively.

Project managers will also benefit by understanding the psychology and sociology of the team through the use of personality indicators or group dynamics. Focusing on the well being of the team and their internal relationships often results in a significant increase in productivity. Effective project managers remember that team members make or break a successful project and therefore, manage the job satisfaction of the project team on an ongoing basis.

Communications Management

Project communications management focuses on the ability to communicate the project's performance reporting to all stakeholders to keep everyone fully informed of its status.

The communications management process includes several different areas of interest:

Communications planning
Information distribution
Performance reporting
Stakeholder management

Communications planning includes the development of a communications management plan. This includes the stakeholder communication requirements, the information to be communicated, and the frequency of communication. It is intended to ensure that the right information gets to the right people at the right time in a format that will be effective in transmitting the status of the project.

This is of crucial importance to the effective project manager, as a lack of communications is often perceived as a lack of control over the project resources. Too often, project managers believe that as long as they do not communicate any negative information, the stakeholders will assume the project is going well. In fact, the stakeholders will often assume the opposite to be the case. Therefore, it is in the best interest of the project manager to communicate effectively and frequently.

Additionally, it is of importance to the project manager to ensure that negative information is transmitted immediately to the interested stakeholders. Often we feel justified in delaying or eliminating negative information from our communications, but this leads to a lack of credibility and trust.

The mechanisms chosen for communication often have a distinct impact on their quality. As much as possible, we should focus on face-to-face communication, as this produces the most effective transmission of intent to the receiver. This is due to the fact that only a small portion of the communication is based upon the contents of the words, whereas the vast majority of information is communicated through body language and the tone of communication. To ensure that the majority of information is effectively transmitted, then, we must seek out face-to-face communications when possible.

Project managers must also be aware that individual differences will have a distinct impact on communication needs for their team members. Introverts are more likely to utilize private communications, while extroverts seek out face-to-face interactions. Intuitive people will more readily perceive patterns and strategic issues, while sensing people need step-by-step instructions. Thinkers seek to understand the sequential logic behind conclusions that have been drawn, while feeling people are much more interested in the emotional content of the message. Judging people are effective at setting and meeting deadlines, while perceiving people often need more assistance in keeping to the project plan. If a project manager is aware of these challenges, he or she will more readily construct communication in a manner that will enhance the transmission of the message.

Performance reporting is a must on any development project. It takes the form of weekly status reports from individual team members that provide detail of what has been accomplished, what will be accomplished during the next week and any concerns or issues that have arisen on the project. These reports are valuable to any stakeholder seeking a status update and to the individuals performing on the project alike, as they require a collective report of accomplishments and concerns at the end of each week. It does not take very long for individuals performing on the project to recognize that if a written report is required at the end of the week, they might want to make sure that some accomplishments are achieved in order to report them at the end of the performance period.

In addition to status reports, progress reports might also be provided on a regular basis. Progress reports differ from status reports in that they usually cover a longer span of time as well as details of progress at a much higher level; they are a summarization of status reports. While status reports are written by all project team members, the project manager develops a single progress report for the entire project team. This necessarily requires the level of detail to be reduced when compared to the individual status reports. The progress report is also a good method for providing the forecasts of the outcome of the project to its stakeholders.

Finally, the project manager must be constantly aware of the satisfaction of the stakeholders. To accomplish this effectively, he or she must communicate with them on a regular basis to ensure that all are satisfied with the current status of the project. If this is not done, project managers will likely receive a great deal of 'help' from their sponsors or upper management, and most do not welcome the increased level of attention.

Project managers will generally benefit by creating templates for these various communication needs, as well as schedules for the communication expectations. By providing these documents and schedules to their team members, they can ensure that an appropriate business rhythm is developed and that regular communications become an expected task.

Risk Management

Risk management is a methodology utilized to identify, analyze, and respond to risk throughout the life of the project in order to maximize the probability of its success. Risk is often minimized in projects, but it can help provide a high degree of insight into areas of concern, enabling the project manager to better understand appropriate project scope and realistic estimates.

Research suggests that risk is a major shortcoming on most project teams. In fact, a recent study found fifty-five percent of runaway projects, those projects that have significant budget and schedule overruns, did no risk management at all.

Risk is defined as the probability of an event occurring that will have a positive or negative impact on the project. In general, a project manager must consider two components of risk: the probability of occurrence and the impact to the project. In a manner similar to how cost control is conducted, impact will be expressed in terms of the monetary impact to the program.

Risk management is a crucial element of any system's development. To understand its methodology, we consider the characteristics of risk and uncertainty, and utilize them to create a rational risk register that describes our development's existing vulnerabilities.

There are several actors of interest for us to understand:

Issues: Anything that requires the project manager's attention
Problems: Anything or an event that has gone wrong and requires action to correct
Hazard: A situation or condition that is likely to cause you harm or loss underserved and circumstances
Crisis: An unplanned event or possible situation that triggers a real or perceived threat to safety, health, the environment or an organization's repetition
Risk: A potential problem or event that may occur and if it does, may cause harm or opportunity

Risk is not a certainty, but it does imply that if a course of action or inaction is taken under conditions of uncertainty, the risk takers may be exposed to possible loss or gain regarding their desired outcomes.

Risk tolerance is the amount of risk that is acceptable to the individual, organization or client. It will certainly be different for all of the individuals working in an organization. This must be considered when assessing risk on any system's development, as when that information is communicated to different parties within the organization it must be tempered with regard to their particular levels of risk tolerance.

Risk: A future event or series of events with a probability of occurrence and the potential for impact to the objectives of the program. This implies that there are two components to assessing risk: probability and the impact to the program.

The risk management methodology contains a five-step program. These steps include:

Risk identification
Qualitative risk analysis
Quantitative risk analysis
Risk response planning
Risk monitoring and control

Risk identification is one of the most challenging aspects of a risk program. It requires the system engineer to consider all of the complexities of the program by asking several questions:

What are the things that could potentially go wrong on this program?
What are the possible causes for those things going wrong?
If things do go wrong, what are the potential consequences?
What are the things that could potentially impact the program positively?

These risks are typically described in a narrative form that defines the specific concern in a manner that will reduce the nebulous nature of the statement to provide specificity to the developer. Risks are logged in a table that will be utilized throughout the life of the system's development.

Risk management monitors and adjusts risks on a systematic basis during the program development cycle. In most cases, this translates to a review of the risk register during regular project meetings, where each risk is assessed to define if any mitigation strategies have been implemented or if occurrences on the project have changed either the probability or impact of any of the risks. The risk register and the corresponding heat map are then updated and communicated during the regular business rhythm within the organization.

Procurement Management

Project procurement management focuses on the ability to manage the external entities that provide the services or materials that will be used on our projects. It can be considered to be a vertical integration strategy for the project manager.

Procurement refers to the acquisition of goods or services from an outside source, and may also be called purchasing or supply-chain. This might include the purchase of tangible goods or services. When a service is purchased, it is usually in the form of a contract, which is a mutually binding agreement that obligates the seller to provide the specified items and services and obligates the buyer to provide an agreed-upon price.

Project procurement management includes the acquisition

of all goods and services for the project, and includes:

Acquisition planning
Contract planning
Proposal requests
Supplier selection
Contract administration
Contract closeout

In acquisition planning, the project's needs are utilized to identify where it would benefit from the purchase of goods or services. If there is no need to purchase services external to the project, there is no further need for the procurement management process.

Different tools and techniques are utilized in order to plan for acquisitions on projects. This might include a make or buy analysis. In this analysis and equation is created to allow a comparison between the purchase option and the internally developed option. This allows a financial assessment that will enable an understanding of whether the item should be purchased or not.

When contracting services from external organizations, there are several types of contracts that are typically used:

Fixed-price – a fixed total price for a well-understood and documented product or service.

Cost reimbursable – a contract that requires payment to the seller for all direct and indirect costs. This can be extremely dangerous for the project manager, as it provides no motivation for the seller to control costs.

Time and material – a combination of both a fixed price and a cost reimbursable contract; it places requirements on the seller and limits the seller's ability to recover all subsequent costs.

Unit Price – similar to paying the price noted on the price tag for the good or service.

On cost reimbursable contracts, there are several different forms that are commonly used. These include:

Cost-plus incentive – the buyer reimburses the supplier for allowable costs plus an incentive bonus

Cost plus fixed fee – the buyer reimburses the supplier for allowable performance cost plus a fixed fee

Cost-plus percentage – the buyer reimburses the supplier for allowable performance costs plus a predetermined percentage of costs

The procurement management plan should describe in detail how the procurement processes will be managed, from developing the purchase documentation through to contract closure. There are typically no set methodologies in place for project managers to utilize, and so they are largely bound by their organization's purchasing requirements.

Proposal requests will often be used when the scope of the work or the complexity of the solution is unknown, such as in a highly complex project like the development of a new stadium or a new solution to a technically challenging problem. They can be very useful when the procurement organization has no idea what the optimal solution might look like, allowing flexibility and freedom on the part of the contracted agency to produce an effective solution. They are typically conducted with a simple request for proposal, containing the following items:

Purpose of the RFP
Contracting organizations background
Basic system requirements
Hardware and software environment
Enterprise environment
Description of the RFP process and expectations
Statement of work
Schedule information
Appendices as required

As proposals are collected, they will be assessed for completeness and ranked in order to provide a supplier selection. It is crucially important at this stage for the project manager to ensure that all supplier selection activities are unbiased and objective. Normally, the evaluation criteria for the proposals must be completed prior to the acceptance of a single proposal. From this evaluation, a supplier is selected and placed under contract.

Contract administration is similar to managing the project within the project manager's organization. He or she must identify the type status of reporting, the contractor must provide progress reports, and the project manager must oversee development to ensure that the contract's requirements are being met. In many cases, the legal requirements of the contract limit the project manager's ability to manage the project directly, but he or she must make sure that the contract is completed as planned. As might be expected, the skills required to do so are identical to those needed to complete projects as planned within the organization.

Contract close-out involves the completion of the legal requirements of the contract. It requires the project manager to ensure that all of the requirements of the contract have been fulfilled correctly and satisfactorily, and that appropriate documentation has been provided to the contracting organization.

The Nine Knowledge Areas

The Project Management Institute (PMI) has defined nine knowledge areas that encompass effective project management processes. There are many different interpretations and perspectives on the contents for each of these particular knowledge areas; however, general agreement exists on the basic content of interest. It may be found in *The Project Management Book of Knowledge* published by PMI if the reader seeks additional information.

When considering the skills necessary to accomplish projects in a practical manner, it seems apparent that there are some areas of development for the project manager that are usually not included in the nine knowledge areas. While some argument can be made that this additional information might fall within one of these areas, it is not included in typical approaches and documentation relating to the knowledge areas.

For this reason, a model has been developed that encompasses additional knowledge areas of interest to the project manager. When taken as a whole, it provides a more complete picture of the skills and capabilities necessary for successful project management. A tool is also provided to enable any organization to measure its current project management culture and identify areas of weakness or concern that will in turn allow the identification of improvement activities for its project management domain. To accomplish this, we will begin with the competing values framework.

The Competing Values Framework

Project management within an organization is intended to enable the efficient application of resources to resolve complex problems. We have a tendency within our organizations to spend our time focusing on relatively small gains, with companywide programs focused upon:

Six Sigma
Lean Manufacturing
Total Quality Management
Etcetera

A large portion of the real waste that occurs in our organizations is due to our lack of desire or motivation to correct our projects' shortcomings. These deficiencies cause disruption in the organization that extends well beyond the waste caused by a single project and failures in project execution. Unfortunately, there is no apparent silver bullet to resolve these chronic industry-wide problems.

If we choose to address these fundamental changes within our organizations, we must change our belief systems. This can be done through implementing the Truths of Management:

1. It depends
2. Do what you said you would do
3. Tell the truth
4. You get what you reward
5. Enable others to make the right decision

We also need to understand and embrace Chaos as an

intrinsic characteristic of the universe, knowing that a system allowed to self-organize will arrange itself into a state of critical mass. A system at critical mass is:

Inherently unstable
Susceptible to catastrophe with any small disturbance
Aperiodic
Ergodic
Predictable
Unpredictable

Our organizations are both organized and self-organized. Additionally, our organizations are studies in disturbances:

A typical employee is interrupted every nine minutes
Executives are interrupted every three minutes

Once we have embraced these needs in our organizations, we may address them by electively changing our behavior through examining the coherence of our outlook, while assuring consistency and rigor in our unified approach to project and personnel management. To accomplish this, we must conduct an assessment of cultural limitations, creating a resolution to address the needed changes within the limits of our organization and culture. As can be seen in Figure 77, our organizations are constantly focused on competing needs and requirements. These include needs provided by the customer, advantages and limitations introduced by the culture, and the inherent capabilities provided by our organization.

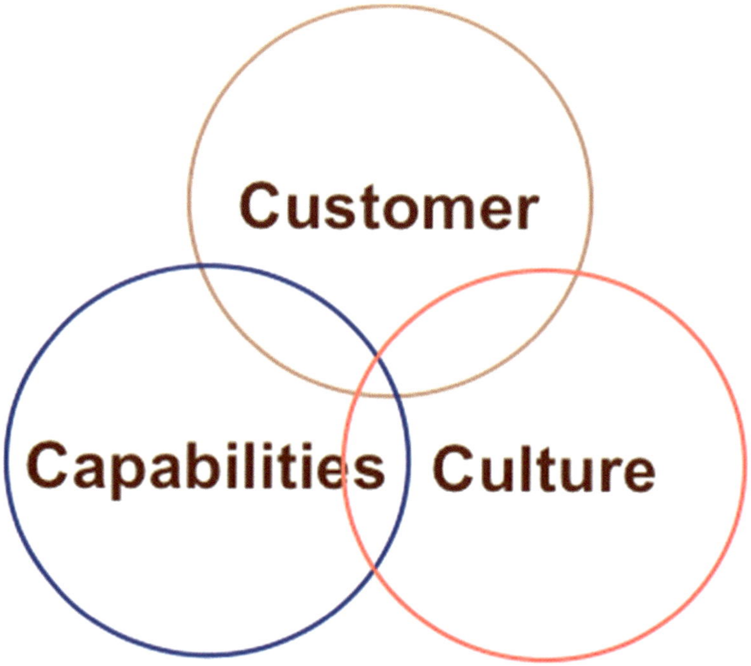

Figure 77: Leading Cultural Change

Any change we wish to implement in the organization must be coherent with our customer, capabilities and cultures simultaneously. If we focus for a moment on how and why work is done the way it is today, we find our workplace to be a confusing and dynamic place. And yet, there is value in being able to understand the hows and whys for our workplace decisions.

Much in the same way we found value in seeking a taxonomy to understand a personality or a social style, methodologies exist for us to assess and describe the culture of our organization. This quantification allows us to begin discussions about our culture and reach a common understanding of its description.

One way that is used to describe culture is:

The way we do things around here

If you were going to describe the way you do things, as if your culture was a person in the next room, how would you describe or define that culture?

Once we have defined our culture appropriately, we can utilize this tool to enable us to identify areas desired change.

But what changes are we talking about?
Are we talking about changes in individual behaviors?
Changes in personnel?
Situational changes?

We know that focusing our attention on both the actions and attitudes of the individual can impact personal behaviors. By observing people's actions or behaviors, we can adjust their tasks to better fit their knowledge or aptitudes. With regard to motivation, we need to pay attention to individuals' needs for affiliation, power, achievement and security, as well as to their personality traits and social styles.

If we choose to focus on change in the situation or organization, we might choose to focus upon changes in our:

Vision
Values
Strategy/Mission
Tactical influences

The tactical influence systems present in most organizations include:

Leadership
Communication
Structure
Policies
Physical layout
Environment
Staffing personnel selection
Ceremonies and events
Goals and measures
Rewards
Information and knowledge
Training and competency development

These are all areas where we may choose to make changes to our methods and processes that will result in significant changes to our organizational culture. Therefore, these are the leverage points by which we can impact our culture through mechanistic changes in our organizational approaches.

We will begin by developing a method by which we can map our organizational culture. An appropriate methodology is found in the competing values model.

The model is constructed of two dichotomous factors that act independently with regard to describing the culture. For this reason, we can place these two factors on an orthogonal grid, enabling us to create a two-dimensional image describing our organizational culture.

The first factor deals with the general direction of focus within the organization. This can be externally focused, outside the organization or outside of the specific project, or internally focused, more concerned with internal factors then external.

The second independent factor of interest is the direction of control within the organization. Does the organization seek flexibility and adaptability, or does it generally move towards increasing levels of stability and control?

When we place these two independent factors orthogonally into a diagram, we begin to understand the value of the competing values framework. We can see in Figure 78 at least four potential organizational perspectives that may be embraced by the organizational culture of interest.

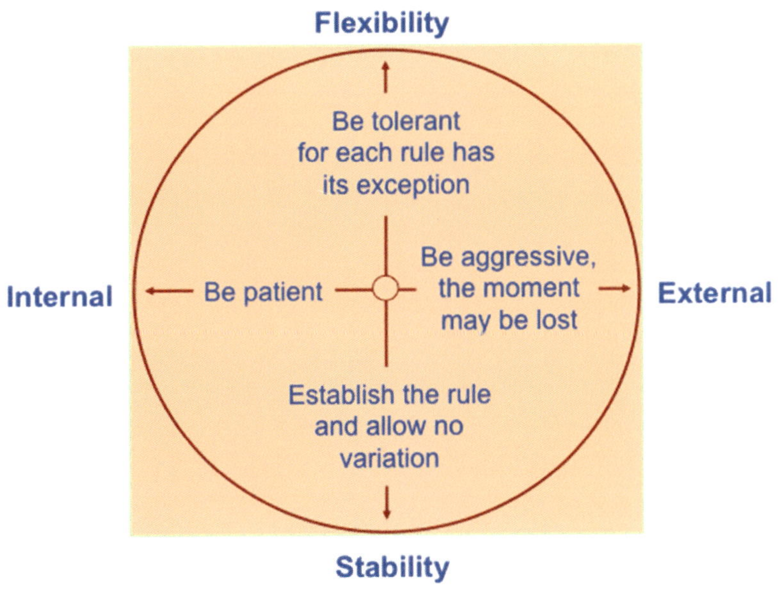

Figure 78: Competing Values

Organizations that seek stability and control establish significant and stringent processes for accomplishing organizational tasks. This is based on the presumption that if the process is adequately defined and control is exhibited within the organization, end results are highly predictable.

Organizations that seek flexibility and adaptability will allow much greater leeway amongst their employees to make decisions, allowing them to deviate from existing processes and controls in order to gain advances that might not otherwise be possible.

Internally-focused organizations will typically demonstrate a significant interest in advancing the capabilities and processes within the organization, establishing completed tasks and schedules for the good of the schedules that exist within the organization as opposed to any specific external needs that might be communicated.

Organizations that have a significant external focus will spend a great deal of time communicating with the stakeholders and customers on the project, ensuring that there is a full and complete understanding of the customers' and stakeholders' needs prior to addressing any work that must be accomplished.

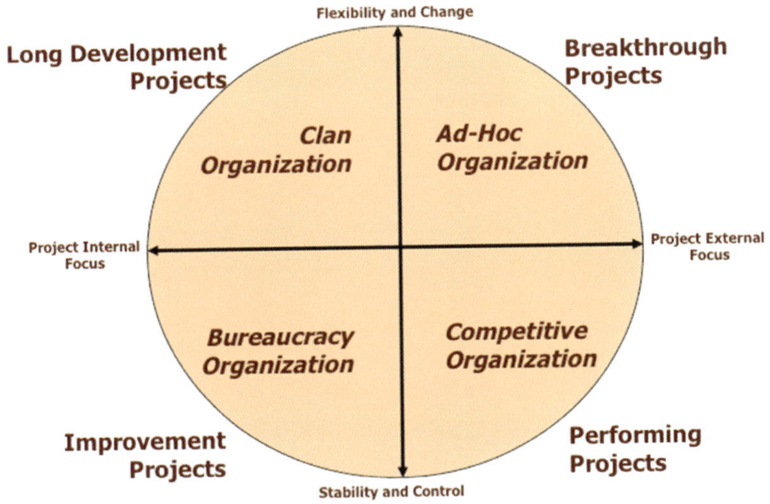

Figure 79: Competing Values Framework

This can obviously result in a description of organizations within each of the quadrants created by the competing values framework. This framework enables us to identify on a macro basis the specific generalized organizational interest areas, as can be seen in Figure 79.

This leads to the definition of four basic culture types that include:

Clan culture
Ad hoc culture
Bureaucracy culture
Competitive culture

The clan culture organization is considered to be people-oriented. It is typically a friendly place to work where people share a lot about themselves; much like might be seen in an extended family. The leaders are mentors and facilitators;

building teams and developing people is important to the organizational leaders. Success is defined as being derived from caring about the customers and a concern for the individuals operating within the organization. The basic tenets of the organization are focused on loyalty and tradition. Job satisfaction and commitment is very high. The importance of developing the human resources within the organization is emphasized, along with high cohesion and morale amongst the participants. Management places a premium on teamwork, collaboration and training.

The ad hoc culture is dynamic, innovative and creative. Entrepreneurial in nature, its people are willing to take risks in order to achieve new things. Innovation and alternative ways of thinking characterize the leaders. Success is defined as having one-of-a-kind or first-to-the-market type products or services. Finding new ways of accomplishing things and new ways of providing value to the customer is the main goal. The tenets most important to the organization are commitment to being on the leading edge of development, business growth and expansion of the business, and acquiring new capabilities and resources. Change and readiness to meet new challenges are stressed, and internal chaos is typically very high. The leaders are innovators and integrators. Individual initiative, innovation, creativity and individual freedom are emphasized.

An organization with a bureaucracy culture features a formalized and structured work environment focused on establishing processes for all functions. What its people do is typically governed by rules and procedures. The leaders are organizers and efficiency experts. Dependable, on-time delivery, smooth scheduling and the meeting of budgets are critical. Maintaining a smooth running organization with few surprises is very important and the main tenets of the organization focus on the formal rules, policies and standard processes. Long-term stability and predictability are emphasized. The managers are coordinators and supervisors. Leadership's main concerns are focused on stability, reliability, dependability and predictability.

The competitive culture is goal-oriented, rational, measurement-based and provides a competitive place to work. In this culture, driven people are primarily concerned with getting the job accomplished. The main focus is on mission success. The managers are producers and directors; they are hard driving, action-oriented and focus on quick decisions. The tenets of the organization focus on winning, goal achievement and building of competitive strengths. Competitive actions, achieving all objectives and working long and hard is highly valued. Being competitive, winning and "being the best" are the key areas of focus for the leaders within this type of organization.

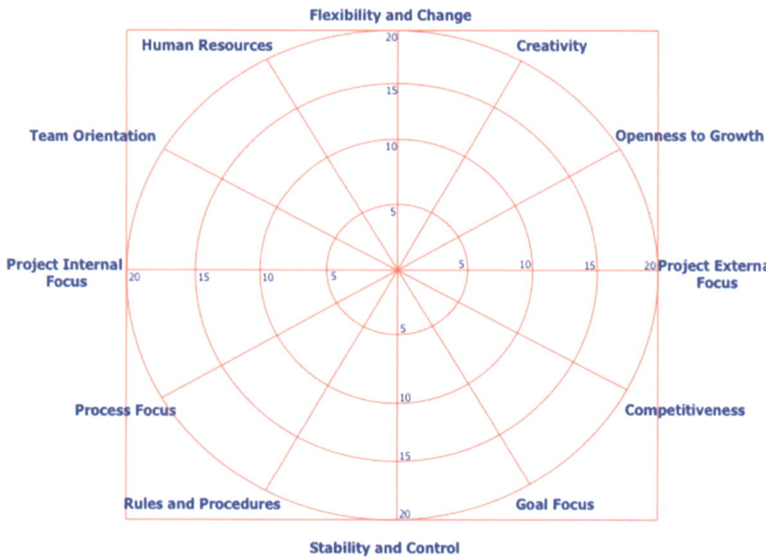

Figure 80: Competing Values Model

The competing values model can be utilized to assess and understand an organizational culture, but it requires that some additional resolution be added to the model, as shown in Figure 80. As can be seen, this model incorporates two distinct measures for each of the four quadrants resulting in an eight-factor analysis of dichotomous terms to enable a mapping of the organizational culture. This is a crucial observation, as it is important to measure both independent factors simultaneously in order to get an effective picture of the culture within the organization.

To accomplish this task effectively, a survey was created that was completed by members of the organization who were very familiar with its cultural behaviors and tribal customs. It was designed to be completed in a short period of time, but allows reasonable resolution within the measures to create notable differences in the cultural descriptors. The survey is included in the following paragraphs.

Project Culture Assessment Sheet

Place an X in the box that best represents your view of the current Project Culture at your organization. Please use the following convention:

1 – Not like our organization at all
2 – Slightly describes our organization
3 – Somewhat like our organization
4 – Very like our organization
5 – Describes our organization very well

Question	1	2	3	4	5
Creativity					
People really listen to different problem solution suggestions					
New ideas are encouraged and embraced					
We recognize and reward risk takers					
It is okay if some of the high risk projects don't succeed					
I would prefer to run a different project rather than run one I know a great deal about					
Openness to Growth					
People like to act on their own initiative					

We react well to surprises in our projects, especially when we discover something new					
Employees have the authority to act independently					
We find opportunities where other people might find problems					
The way we manage projects is constantly changing and improving when we learn new things					
Competitiveness					
We know about our competition and we do everything we can to make our projects more attractive to customers					
We can be ruthless in our desire to complete our projects effectively and efficiently					
We constantly try to understand our customer's needs and adjust our projects accordingly					
When managing our projects, the successful completion of the plan is crucial to us					
Our customers know they can always rely on us to exactly what we promised to do					
Goal Focus					
Once we decide the plan we intend to follow, it is very difficult to change the direction					
Sometimes it is more important to finish the project than it is to adjust our plans because of employee need					

Plans are created to be used as roadmaps; when we follow them we get good results					
Sometimes people get in the way of our organization's completing the work that simply must be done					
If the customer wants it done, then we will do it, regardless of the impact internally					
Rules and Procedures					
When people don't follow the rules, we all pay the price					
We have discovered the best way to do things through trial and error, we need to utilize this past experience					
We don't continue to make the same mistakes, because we document the right way to do things and follow the rules					
There is a clear and systematic method for everything; we just need discipline to be able to stay on course					
It is important that people understand the way we do things, otherwise they won't be able to perform well					
Process Focus					
There is a correct way to do our projects, and a wrong way to do them. We need consistency from all our employees					
We understand the need to define metrics and measure progress, as it helps us to get the job done					

We don't like "loose cannons" in the organization causing all kinds of problems that impact our projects					
When people act independently they mess up the smooth running machinery					
We work best when things remain the same; it's the needless change that continually messes us up					
Team Orientation					
Consensus creates a more accepting atmosphere, and we recognize the value of achieving it					
Without our collaboration we would be unable to complete our work efficiently					
We accomplish much more when we allow people the freedom to collaborate and communicate					
We like to recognize and reward our teams as a group rather than as individuals					
People would prefer to be part of a team rather than stick their neck out alone					

Human Resources					
Choosing between the project and the people is easy; the people come first					
We develop our people by giving them stretch assignments; it is a necessary cost to the business					

People are more than just a commodity; we focus on them and try to do the right things for them					
If we need to miss a delivery to accommodate an employee, we would be willing to do so					
It just isn't smart to be hardnosed about the project when our people have outside issues to deal with					

Total up each of the focus areas according to the point assignment on the first page. Then place the total from each of the focus areas in the boxes below:

Creativity	
Openness to Growth	
Competitiveness	
Goal Focus	
Rules and Procedures	
Process Focus	
Team Orientation	
Human Resources	

Upon completion of the survey, the individual may summarize his or her findings on the last page. This will enable the survey taker to map his or her view of the corporate culture utilizing the previous diagram for reference. The measures selected for use in the survey were identified as crucial aspects of the competing values framework, isolating specific organizational patterns that might be identifiable and of interest to the organization. These patterns are organized in a manner that is somewhat dichotomous along the axes of interest, although it is clear that to excel in one dimension, it is not necessary to be deficient in the opposing direction.

The four cultural types are described by generalized characteristics in Figure 81. This chart provides us some insight as to the demeanor of the organization and the overall philosophy that is typically followed when pursuing projects.

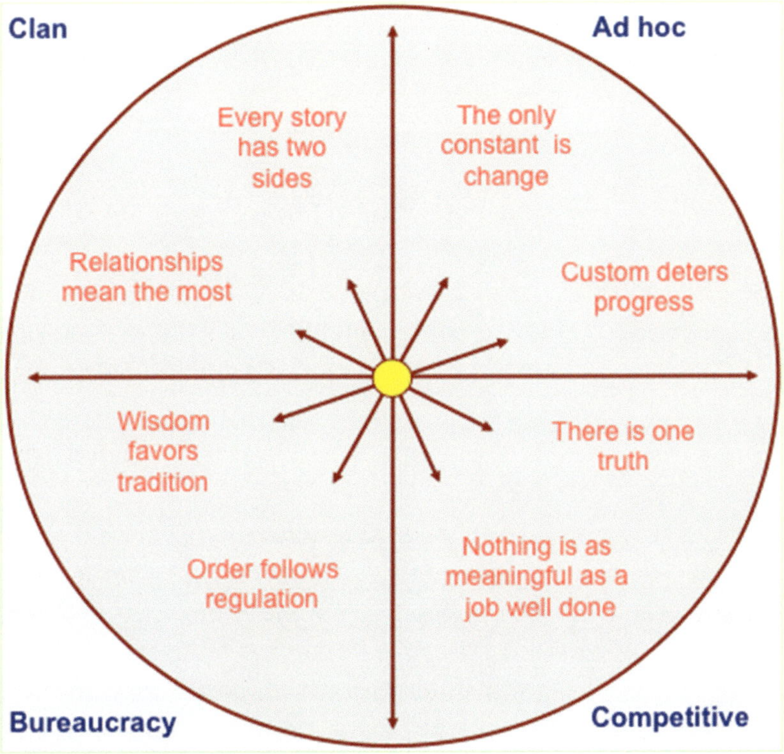

Figure 81: Competing Values

Upon reflection, it should be clear to the reader that any organization shares characteristics from each of these quadrants. After each of the factors is calculated, it can be graphed on a chart, resulting in a shape that describes the organization based upon the competing values framework definitions. One is tempted to assume that each of these factors should be held in some sort of balance, thereby creating an optimal or Utopian organization. While in some cases this might be desirable, in most cases the organization and industry dictate that a different shape is preferable in order to remain competitive in that industry.

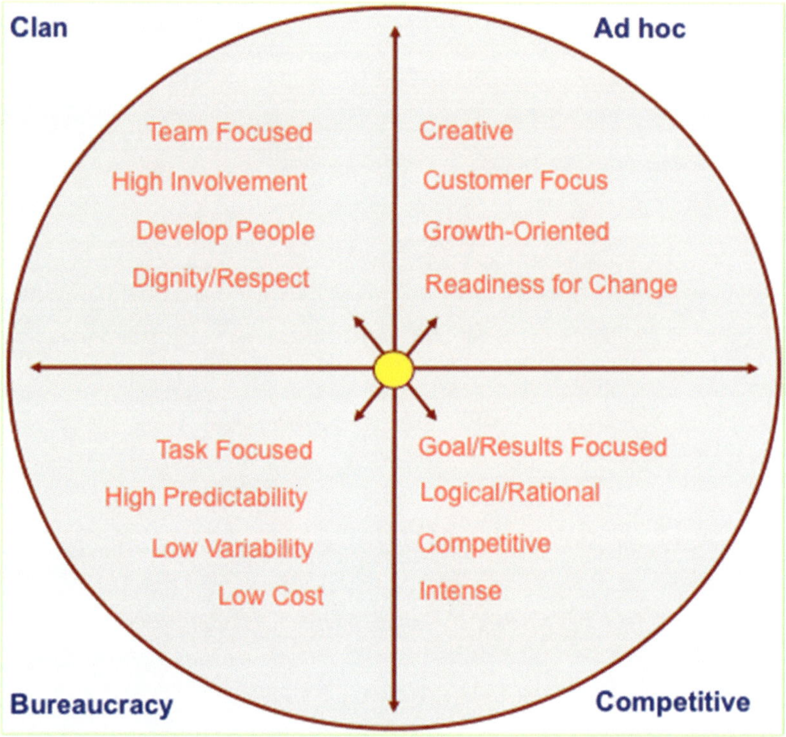

Figure 82: Competing Values Characteristics

In reality, although there are specific questions that are associated with each of the eight factors interest, there is a natural smoothing that occurs in the graphing of the organizational culture. This is largely brought about by a lack of direct questions along each of the orthogonal axes, as well as the perspective that the total area included within the shape for each of the quadrants provides some measure of the strength of that cultural type as a descriptor of the organization. In Figure 82, additional descriptors are provided to further clarify what areas of interest are typically exhibited within a cultural type.

If an individual would take this survey with respect to his

personal preferences with regard to projects, he could assess personal project cultural preferences and clarify the specific type of project leader he is likely to resemble when managing a project in his ideal world. These individual archetypes might be useful for the reader to identify areas of strength or weakness with regard to his or her approach to project management. In order to identify your personal preferences, after completion of the survey and mapping of your preferred corporate culture, you may identify which area or areas demonstrate the highest scores, thereby indicating the areas of greatest preference. The archetypes are shown in Figure 82.

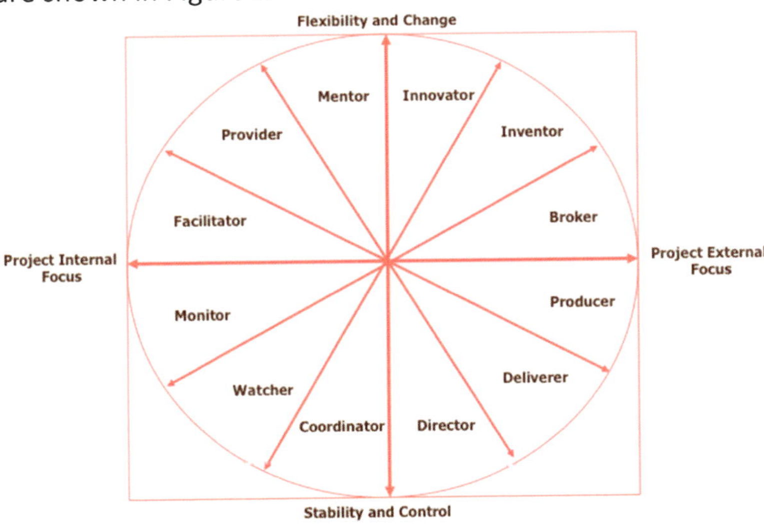

Figure 83: Leader Types in the Competing Values Framework

Each of the cultural archetypes has interesting characteristics that may be used to identify personal strengths and weaknesses with regard to project management. They can help identify how your project team might perceive you as the project manager, thereby clarifying to some degree the relationship that can be expected between you and your project team. We will discuss each of these archetypes individually.

Facilitator

The facilitator excels at generating collective efforts that are built upon completion in teamwork, and will likely minimize interpersonal conflict. In this role, the leader is described as process-oriented. Expected behaviors will include effective mediation, effective conflict resolution techniques, developing cohesion and positive motivation, generating high productivity and participation, and facilitation within the group environment.

Provider

The provider is expected to have his or her project team's backs. This individual is particularly effective at producing loyalty and engagement on the team, and will likely become the person that most team members feel they can trust to do the right thing with regard to the people in the organization. This person will demonstrate the behaviors of effective negotiations, communications, emotional intelligence, a high regard for others, and respectful behavior towards others. In this role, a leader is typically described as a caring person. Expected behaviors will include a focus on the person, and not on the resources consumed by the project.

Mentor

The mentor is identified by his or her engagement in the development of individuals who are contributing to the project team. This is typically a high empathetic orientation, and the leader will likely be helpful, approachable, considerate, a good communicator, and someone who will go to great lengths to demonstrate fairness. In this role, the manager will be an effective listener, and will address the concerns of the team through appropriate negotiations. This manager will also typically demonstrate a great deal of positive feedback for the team, and will become known as a leader who creates development plans for the growth of team members, as well as for him or herself. The mentor will demonstrate skill building, cross training, opportunity generation and mentoring.

Innovator

As an innovator, a manager is likely to accept a great deal of adaptation and change. This manager will thrive on a chaotic environment, understanding the need for a messy system until an appropriate time at which he or she will be able to generate order and produce the deliverables for the project. This manager will be viewed as a highly creative and integrative individual, with visionary tendencies and the ability to engage to team effectively in the project development. Typical behaviors include brainstorming skills, communication skills, collaboration skills, care for the team, and an ability to effectively understand the needs and desires of the team in order to create engagement on the project.

Inventor

The inventor is expected to generate out-of-the-box thinking and vision reaching well into the future. This manager will excel with vague schedules and nebulous requirements, thriving on the chaos generated by the ambiguity. This requires behaviors that include a high tolerance for requirements volatility, delegation skills, a high degree of trust in the team, and comfort with incomplete or limited communications. The inventor will typically value the ability of team members to approach things from different perspectives, and to generate multiple solutions to a perceived problem. This type of manager will demonstrate the ability to persuade others of the merits of his or her approaches, to produce effective negotiations, to engage in effective people reading and to be adaptive in a highly dynamic environment.

Broker

The broker is particularly effective at managing all of the external interfaces for the project team and obtaining adequate resources to accomplish the project's deliverables. This manager demonstrates a high degree of political skill, along with the ability to effectively use rhetoric, persuasion, and influence management, and will be effective at the application of power. The image and reputation of the team are important to this manager, and it is likely that he or she will quickly respond to challenges in this area. This manager will spend much time dealing with people from outside the project team, and will represent, negotiate, sell, and in general, act as a liaison and spokesperson for the project team.

Producer

The producer is a task-oriented and deliverable-focused individual who possesses a high interest level, motivation, energy and personal drive. This manager will be highly effective at working in autonomous environments, and will encourage his or her subordinates to act in much the same way. He or she will expect his or her subordinates to accept delegation, take responsibility, complete their assignments on time and budget, and maintain a high productivity. The skills required to accomplish this are effective communication skills, personal motivation, conflict resolution skills, and an ability to model the type of behavior expected.

Deliverer

The deliverer is recognized by the customer as the person who can be relied on to make sure the job is done properly. The leader in this role will be deliverables-driven, highly motivated, and willing to go the extra mile in order to make sure that all commitments are delivered upon. This individual takes promises very seriously, and considers it a personal failure if any of the project deliverables are not met. He or she will feel a strong connection to the customer and the competitive advantage realized by the project. The leader in this role will typically demonstrate characteristics of a driver, and will expect a similar approach to accomplishing work in many of the team members. Finally, the deliverer will continually focus on accomplishing the right things in order to make the promised deliveries.

Director

A director will attempt to clarify expectations through managed processes and formats, with highly detailed planning and the definition of appropriate requirements and deliverables. This leader is a highly autonomous contributor with well-defined problem-solving skills, trade-off analysis skills, project planning skills, role definition skills and tasking skills, and will typically adhere to highly consistent rules and policies regardless of the project. There is typically a high degree of acceptance on the part of the project team, as few surprises are introduced later in the project.

Coordinator

A coordinator is identified by his or her need to maintain appropriate structure and flow within the system. This manager will take pride in being particularly dependable and reliable with regard to the completion of project objectives. He or she will focus on the appropriate methodologies that should be followed by the project team, and will advance processes in a manner that will enable more efficient and effective use of the project's resources. Typical behaviors include change management skills, conflict resolution skills, administrative and communication skills, and an innate ability to status the project rapidly and effectively, thereby reducing the impact of outsiders on the project team. This type of manager will also be highly effective at producing project plans that are achievable and objective.

Watcher

The watcher feels he or she must be aware of the performance of all individuals, as well as of the resources within his or her area of responsibility. This manager is highly aware of the various processes and policies that apply to any given situation, and possesses a need to ensure that everyone adheres to those requirements. He or she is under the impression that if we simply got the process right, everything should run effectively under all conditions, and we would all be better off and accomplish much more if we were to follow the process. The leader in this role typically values predictable and reliable behaviors and performance on the project team. Typical behaviors include a focus on processes and planning, adherence to project management methodologies, an ability to accept deviations from expected paths, and a general preference to focus on process rather than the people on the project team.

Monitor

The monitor wants to know everything that is going on in the project at any given time, and uses this information to establish whether the individuals operating on the project team are compliant with all of the rules and procedures within the organization. This type of leader is less interested in the specific problems and challenges the team is facing, and more focused on whether the members of the team are meeting their quotas. The monitor typically possesses a passion for details, and is very effective at trade-off analysis. Behaviors in this role include technical analysis, understanding data-driven decisions, project planning, adherence to policy, and effective status reporting and communication skills with regard to organizational leadership.

Understanding these archetypes allows us to assess the areas of the project management domain we find most comfortable and that are supported by our strengths. By understanding our predispositions, we understand the types of companies wherein we may excel as project managers. Additionally, we can consider if the culture defined within the organization is particularly effective at accomplishing our projects' objectives. Therefore, this tool can be utilized as a problem-solving or continuous improvement tool within organizations in order to gain insight into what is working and what is not.

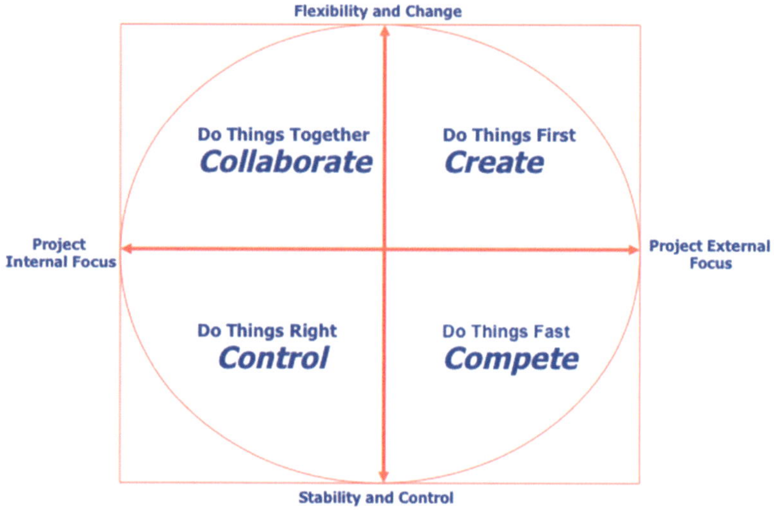

Figure 84: Competing Values behaviors

In Figure 84, we see the preferred manner in which things are accomplished for each of these four cultural types.

The clan culture prefers to collaborate overall, doing things as a cohesive unit.

The ad hoc culture focuses on being first to the finish line, doing things before anyone else is capable of doing them. The bureaucracy culture focuses on following processes and procedures, accomplishing things in the ways that are expected.

In the competitive organization, Individuals are expected to do things quickly and effectively in order to accomplish the objectives outlined by the customer. The customer is the most important thing.

Measuring Corporate Culture

The competing values framework has been used effectively to identify the challenges faced by cultural-based organizations. In general, these problems are deeply rooted in the past, and are perceived as limiting to the organization's ability to compete going forward. This tool was developed to enable organizations not only to identify their own cultures, but also to identify the areas of development they seek to focus on to increase their capabilities.

An example is provided from an organization that utilized this approach in order to identify areas of focus for change. This was a large organization, privately held, whose leadership understood that in order to begin to institute cultural change, it was necessary to identify specific tasks for implementation.

To begin, the previously described survey must be supplied to a large group of individuals familiar with the organization. In most cases, it is expected that this team will be comprised of at least ten individuals, and can be effectively managed in groups of up to one hundred twenty. The steps in the process to generate a common corporate culture are as follows:

1. Provide a survey to each individual participating.
2. Each individual completes the survey without discussion within the group.

3. Once all surveys have been completed, individuals will graph their corporate culture shape on the competing values framework is shown in Figure 80.
4. Participating members will pair up with another individual and begin to discuss their corporate culture assessment shapes with one another.
5. The teams of two will discuss areas of deviation between their corporate culture shapes, establishing interpretations and reasoning in order to reach a consensus on a common shape.
6. Once a common shape has been agreed upon, each team will pair with another team and repeat the process.
7. This process is continued until the entire group has produced a common corporate shape that they believe describes the culture within the organization of interest.

This will generate a shape such as shown in Figure 85, representing the company's current state. Of particular value in this process is the large amount of discussion that will occur between different members of the teams, clarifying interests, understandings, stories and beliefs within the organization. Most companies that have utilized this approach find the large-scale discussions to be of the most significant importance in understanding their corporate culture. Normally, these discussions include a rereading of the questions and discussions involving the reasoning behind the ratings as provided. This effort typically hashes out any inconsistencies or interpretations that are uncommon within the group.

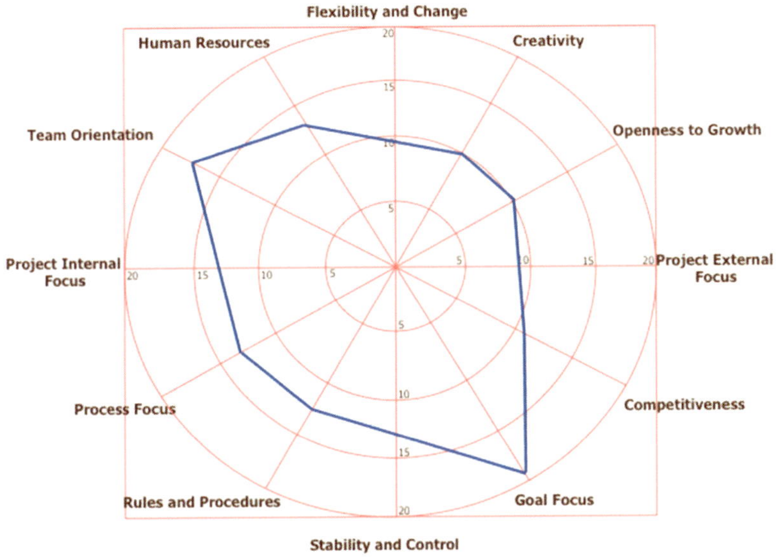

Figure 85: Company's Current State

In the next step, the group members must resolve two simple questions:

We understand what shape our corporate culture is, but what shape should it be? Why?

This will entail a discussion period equivalent to that of the first discussion, establishing a common understanding of what shape would be most appropriate for the organization to effectively accomplish its goals. This is generated over the company's current state, establishing the areas to increase and decrease on the diagram. It identifies areas in which we must define what we are going to start to do to increase in that area, as well as areas where we will define what we are going to stop doing to decrease in those areas. This will generate for us a practical action list that will enable the start of corporate culture change.

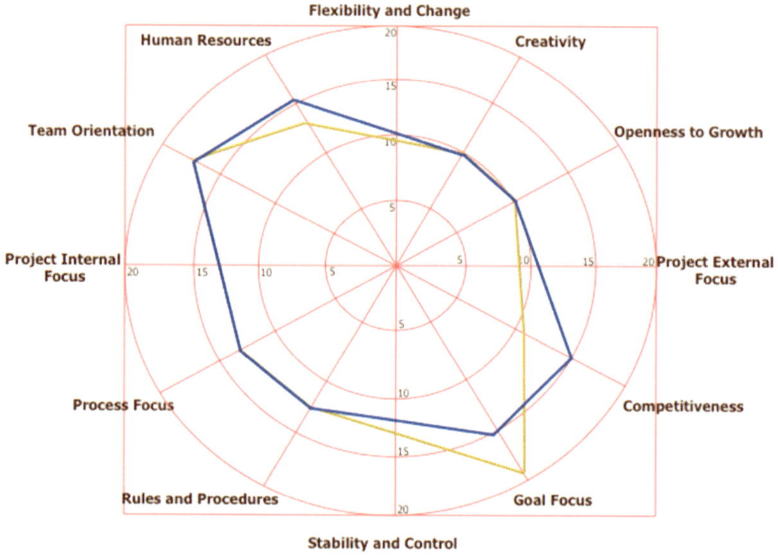

Figure 86: Company's Desired State

Figure 86 is a completed cultural assessment and identifies the areas where movement is desired. In this particular case, the participants established three areas of desired change with respect to their current organizational culture:

Develop greater human resources focus
Develop greater competitiveness focus
Reduce goal focus to a more reasonable level

This exercise allowed the team to focus on smaller components than culture in order to define actionable items. Utilizing their detailed organizational knowledge, they identified specific areas where improvement was needed or where a decrease in focus was required. In this particular case, actions were identified for each of the three areas.

1) **Human Resource Sections**

- Better define roles and responsibilities with appropriate job descriptions and expectations provided to the individuals
- Better business sponsorship
- Have the courage to care about the people
    - Value differences
    - Provide candid feedback
    - Work/life balance
    - Personnel development
    - Establish formal coaching/mentoring of project managers
    - Utilize chaos theory in practice, incorporating 75% utility
    - Allow failure as an appropriate learning step
    - Establish that it is okay to be a leader
    - Staff projects adequately
    - Leadership to stand up for and respectfully support project management processes and procedures
    - Provide safe havens in which project managers may learn
- Establish a training program for project managers in practical skills, including estimating, risk management, conflict management and negotiations

2) **Competitiveness Actions**

- Ensure that the project is worth doing at the pace we are doing it, and not just driving the team to meet an arbitrary date and budget
- Make realistic promises; tell the truth

- More active business sponsorship
    - Find the right sponsors
    - End-user involvement
    - Accept accountability for project sign-offs and sustainment of resources
    - Project manager and sponsor to periodically reevaluate project objectives and benefits to assure validity of project efforts
    - Align the rewards to appropriate behaviors
    - Time individual rewards to company rewards
    - Focus on what needs to be done not how it is done
    - Let people push the boundaries
    - A lot of flexibility dependent upon project content
    - Create a plan, and allow the plan to be dynamic– have a process to refine the plan
    - Capture and measure company benefits and potential impacts from all projects

3) **Goal Focus Actions**

- Cancel projects when appropriate
- Get sponsors more involved; make them accountable for the project and set realistic goals
- Define the sponsor's role in detail
- Project managers to hold the sponsors accountable
    - Train and educate sponsors
    - Expect agility in execution, people and plan
    - Revisit project success measures; what is project success?

Once all of these desired actions have been defined for each of the focus areas, it is clear that not all suggested actions are to be effectively undertaken simultaneously. At this point, it is appropriate to consider a prioritization of the actions defined by clarifying and classifying those included in the list. Once this is completed, the top focus actions can be identified for the organization.

Top Focus Areas

Business and department sponsorship needs improvement – they need to begin to manage outward
Project manager expectations; what are the roles and responsibilities?
Identify, acquire, and train appropriate tools for project managers – Chaos theory, courage, risk, estimating, mentoring, leadership
Define, measure and evaluate project success

This process has been highly successful at moving a large organization from the realization that a cultural change is needed to the identification of tangible actions that might be undertaken in the short-term to begin that change. In this particular organization, once the listed actions were completed, the general consensus on the team was that significant improvements had been accomplished in a relatively short period of time. After these actions were completed, a similar tool identified additional actions, allowing a prioritization and focus to be sustained for continuous improvement.

Once again, it must be stressed that there is no single shape that is considered to be more desirable than another with regard to organizational culture. Each of these factors of interest have been independently defined, and there is no reason to suspect that each of the axes will be normalized with regard to any other axis. Secondly, there is no research or reason to suspect that these results should be treated as an absolute measure of capability, and the reader should focus on utilizing this information as a relative measure of areas of interest.

Once the normalizing discussion has taken place between all of the members mapping the organization individually, a collective understanding of the organizational culture has been defined, and the map is best utilized as a guide to defining the areas we choose to increase or to reduce our focus level.

A Proposed Project Management Culture Assessment

The competing values methodology is very useful in assessing and understanding an organization's culture. As discussed, it can be very powerful in enabling the generation of specific tasks associated with implementing cultural change. This technique has been used with multiple organizations in order to spearhead these very challenging and difficult efforts.

During research conducted on organizational project management, the competing values framework was utilized to better understand and assess how the corporate culture either supported or detracted from an organization's appropriate project management culture. When this eight-factor model was compared to the knowledge areas defined by the Project Management Institute, it was clear that some expansion of the model was necessary to more capably describe the project management culture. This realization resulted in a careful assessment of a new competing values framework that might be more appropriately applied to better understand the project management culture within an organization.

As previously discussed, the competing values framework is based upon two independent focus areas, the first relating to an internal or external focus for the organization, and the second relating to a people or process focus within the organization. These factors are of significant interest to the project management domain, but some adjustments in these descriptors will result in a more effective project management cultural assessment.

The proposed project management cultural assessment includes two independent factors of interest, and continues to utilize a people versus process focus on one axis, but modifies the other axis to assess the basic focus of the project effort, defining this as either a conservative or an entrepreneurial focus. When the nine knowledge areas were mapped against this enhanced model, it was discovered that the knowledge areas cover a significant portion of the defined factors, but there are also significant gaps in the PMI model. This assessment also led to the conclusion that in order to produce a robust and useful model, more than eight factors would be needed in order to define the entire domain of interest.

A survey was then constructed in order to assess the relative strengths of the organizational culture with regard to each factor of interest. When these are mapped against the nine knowledge areas, this tool can be used to assess the project management capability that can be expected for an organization. However, when additional factors are included, a sixteen-factor model is defined that better describes the entire domain of project management, expanding upon the nine knowledge areas significantly.

The Sixteen Values in the project management cultural assessment are:

Creativity
Project Execution
Opportunity Management
Innovation
Negotiations

Management
Integration Management
Procurement Management
Quality Management
Cost Management
Risk Management
Scope Management
Time/Schedule Management
Teams / Leadership
Communications Management
Human Resource Management

Each of these factors is defined by a series of questions within the cultural assessment survey designed to establish its relative strength within the organization. The survey was limited to five questions per factor area, mainly to allow the survey to be completed in a relatively short period of time. In many cases, when a more extensive survey is conducted, it is commonplace for the individual taking the survey to spend considerable time reflecting on his or her perspective of the organization, where as the survey was intended to provide more of a snapshot of the organization. In general, the snapshot survey will more closely resemble the actual culture within the organization and will provide more useful feedback for the individual.

We will discuss each factor in turn, with a description of areas of interest within each one. Since they appear on a continuous model, some overlap occurs between factors, and it is left to the discretion of the reader to decide if questions can be enhanced with additional organizational terminology when utilizing a survey of this type.

Creativity

This factor is intended to focus upon the divergent thought patterns within the organization. It is expected that an organization that demonstrates a high-risk propensity with openness to new ideas and perspectives would score high in this factor. In addition, this factor intends to measure the desire of the organization to seek out non-obvious solutions to problems. This would demonstrate a learning orientation within the organization, and a willingness to take significant development risks if necessary to accomplish technological breakthroughs. In general, different is considered good in this organization.

Project Execution

This factor focuses on the ability of the project leader to instill a sense of autonomy amongst the members of the project team. Autonomy is useful, as it ensures that the intellectual capacity of the project team is focused on the development effort. While this generates a great deal of freedom, the project leader must ensure that the team remains focused on the end goals, implying that substantial time is invested in communicating them effectively. The project leader in this type of organization allows the project team to define multiple paths to a solution, not requiring that a stringent development plan be followed. This organization will have a significant people focus, and will try to address appropriate work/life balance for the project team. Teamwork is valued significantly.

Opportunity Management

In this type of organization, people attempt to remain

positive with regard to challenges faced by the project team. This might be described as a focus on opportunities as opposed to problems. This will require the project team to be open to new approaches, and will also demonstrate a significant customer focus. In general, the project will be approached from multiple points of view in an effort to enhance the project team's understanding. This type of approach will focus on the needs orientation, and will find the defined path of secondary importance. This type of organization focuses on striking quickly in order to take advantage of opportunities presented.

Innovation

This factor demonstrates the desire for the organization to provide autonomy to the individual contributors, allowing them to make significant decisions without a great deal of oversight. This demonstrates a very decisive posture within the organization, and an acceptance of risk and potential failure within the project. It establishes a significant authority for the project team members and results in significant technological innovations and advances on a relatively short schedule. Note that this factor attempts to measure the ability to innovate rather than to simply create. Innovators produce tangible results, and creativity produces ideas. In general, this factor demonstrates that the organization finds change to be highly desirable, and is accepting of the concept that a different way may yield significant returns.

Negotiations

This factor focuses on the ability of the project team to manage conflicts effectively throughout the development cycle. This will be demonstrated with superior communications, including a documented communications plan and a typical business rhythm that will be followed. This communications discipline results in significant benefit for the entire project team. Negotiation skills are considered valuable, and a demonstrated value would include a desire to produce mutual benefit in the event of conflict. This results in an extremely open and trustworthy organization focused on collaboration between the project team members.

Management

This factor focuses on the desire of the project team to establish an appropriate process to maintain significant consistencies within the project team in order to enhance the team's competitiveness, by ensuring that project waste is minimized. This is demonstrated by the use of metrics in the execution of the project, and a desire to remain on course with minimal deviations. It is coupled with convergent thought processes and a general belief that there is a right way to accomplish the deliverables. This may be construed as a lack of flexibility, but this factor establishes the desire of the project team to execute the plan as written.

Integration Management

This factor can be closely associated with a systems approach to project management. In this culture, planning is prized as a method to define the various unknown actors on the project, and a great deal of time is invested in the development of requirements. The project management process is utilized to ensure that the final deliverables are achieved in a systematic and pragmatic fashion, so the project team will focus their efforts on appropriate scheduling and resource allocation prior to undertaking the development of the project. This will necessarily create a great deal of cohesion on the project, and will in general produce a commonly held understanding of what will be accomplished on the project. This will typically result in a high degree of team engagement during the execution of the effort.

Procurement Management

In procurement management, the focus is on establishing openness with suppliers to enhance their ability to contribute effectively to the project. In addition to this, the project team is open to identifying new and different vendors, and will utilize past vendor performance to make decisions regarding partnering with suppliers. This factor will demonstrate a systematic selection process associated with suppliers, which requires effective and accurate planning. In general, companies that rank high in this factor will seek a high degree of collaboration and support from their supplier base.

Quality Management

Quality management focuses specifically on the reduction of waste and variability. This is intended to create a high degree of consistency in project performance, and results in a highly dependable project team. This dependability is typically manifest in an ability of the project team to reach their desired goals with appropriate schedules and budgets. This factor also demonstrates the importance of the voice of the customer, and an ability to effectively deal with ongoing changes that occur during the life of the project. This change management process is utilized as a mechanism to ensure that the project plan is at all times predictive of the expected outcomes.

Cost Management

The cost management focus is associated with the accurate prediction and the effective tracking and reporting of costs during the project life cycle. This is demonstrated with established cost metrics and a scheduled business rhythm to ensure that the project's status is reviewed on an ongoing basis. The intent is to maintain a constant focus on the project's bottom-line, and to utilize it to make predictions about the likely outcome of the project. These predictions are utilized to adjust the existing project plan to maintain a high probability of project completion.

Risk Management

The risk management factor within the project culture focuses on the ability of the project team to accept uncertainty and change during the execution of the project. In general, companies that score high in this factor have well-established processes utilized to help them better understand the potential liabilities undertaken by the project team with regard to future risks. This implies that the project team is constantly reflecting on potential future events and their effects on the project's success. In general, organizations that utilize effective risk management seek to draw down their risk during the execution of the project, demonstrating a significant desire to both manage and avoid risks.

Scope Management

Scope management focuses on the team's ability to ensure that the project plan describes all of the work to be accomplished on the project, and that it is adjusted accordingly when changes are required. This is accomplished with highly detailed planning processes and significant communication requirements on the project team. Documentation is important, and a formalized change management process will likely be present. The project team in this culture will seek a high degree of coherence, which limits waste and deviations from the project plan. This desire for a higher degree of consistency in the performance of the project can result in a highly adaptable and flexible team, but also ensures that the project documentation reflects reality at any given time.

Time/Schedule Management

This factor is primarily focused on the predictive effectiveness of the scheduling process. The project team that scores high in this factor will take their promises very seriously, and will take great pride in their ability to effectively predict the outcome of the project. This factor goes hand-in-hand with the cost management factor, whereby the project team is highly engaged in tracking and reporting the project's status. These factors are utilized to make predictions with regard to the completion date of the project, and to establish the amount of progress that has been accomplished. Documentation is highly valued, and the ability of the project team to meet the scheduled finish date is of primary importance.

Teams/Leadership

This factor focuses on the value of teamwork. Organizations that score highly in this factor value consensus over autonomy, ensuring that significant collaboration is established within the project team. This results in a high degree of employee engagement in the project's goals, and produces a great deal of synergy on the project team. While these efforts require a greater investment of time in the project developments, the group dynamics will be significantly improved. Project team members recognize that the relationships developed during the project are of significant value to them within the organization. This type of project culture values people over accomplishments, and is based upon the premise that if you take care of the people they will take care of the work.

Communications

The communications factor is intended to measure the importance of ensuring that all stakeholders are informed of all the events on the project team. This requires that a significant business rhythm and communications plan be established early in the project life cycle, ensuring a discipline is established through out the execution phase. This will also be coupled with appropriate project tracking mechanisms allowing effective status updates and accurate prediction of likely outcomes. The individuals associated with this type of project culture value their ability to listen to other members of the project team, paying close attention to both verbal and nonverbal communications. This results in a higher degree of connectedness amongst the project team members and a high degree of loyalty on the project.

Human Resources

This factor focuses primarily on the importance of people in the project's development. This people focus includes an ability to effectively communicate amongst the project team, but extends beyond that to focus on the development of people and a desire to better understand the individuals on the project team and their differences. The project team directs a high degree of effort on establishing fairness within the project team, ensuring that all members of the team maintain job satisfaction during the project's execution. This yields a highly engaged team, and generates a highly adaptable and flexible approach to the project as a result. In general, teams with a high focus on human resources have significantly higher productivity than other teams.

These sixteen factors may be used to establish the relative strengths of the factors within the project management culture for an organization. As has been previously suggested, the reader should consider if the survey is best utilized as printed or modified in a manner that will employ appropriate terminology for the organization to be measured. These descriptors are provided in order to allow the surveyor to better tailor the survey for his or her organization prior to implementation. Keep in mind that these are relative measures, and the scale between factors is of secondary importance. Of primary importance is the perceived strengths of any given factor within the project management culture, and the reflection that is generated by this survey in order to consider and establish what strength would be desirable for each in order to produce a high-performing project management culture.

The Survey

Project Culture Assessment Sheet

Place an X in the box that best represents your view of the current Project Culture at your organization. Please use the following convention:

1 – Not like our organization at all
2 – Slightly describes our organization
3 – Somewhat like our organization
4 – Very like our organization
5 – Almost exactly like our organization

After completing your five questions, total each section in the space provided.

Questions	1	2	3	4	5
Creativity	Section Total =>				
People really listen to different problem solution suggestions					
New ideas are encouraged and embraced					
We recognize and reward risk takers					
It is okay if some of the high risk projects don't succeed					

I would prefer to run a different project rather than run one I know a great deal about					
Questions	1	2	3	4	5
Project Execution	Section Total =>				
People like to act on their own initiative					
We react well to surprises in our projects, especially when we discover something new					
Employees have the authority to act independently					
We find opportunities where other people might find problems					
The way we manage projects is constantly changing and improving when we learn new things					
Questions	1	2	3	4	5
Opportunity Management	Section Total =>				
We know about our competition and we do everything we can to make our projects more attractive to customers					

We can be ruthless in our desire to complete our projects effectively and efficiently					
We constantly try to understand our customer's needs and adjust our projects accordingly					
When managing our projects, the successful completion of the plan is crucial to us					
Our customers know they can always rely on us to exactly what we promised to do					
Questions	1	2	3	4	5
Innovation	Section Total =>				
People like to act on their own initiative					
We react well to surprises in our projects, especially when we discover something new					
Employees have the authority to act independently					
We find opportunities where other people might find problems					
The way we manage projects is constantly changing and improving when we learn new things					

Questions	1	2	3	4	5
Negotiations	Section Total =>				
When people don't follow the rules, we all pay the price					
We have discovered the best way to do things through trial and error, we need to utilize this past experience					
We don't continue to make the same mistakes, because we document the right way to do things and follow the rules					
There is a clear and systematic method for everything, we just need discipline to be able to stay on course					
It is important that people understand the way we do things, otherwise they won't be able to perform well					
Questions	1	2	3	4	5
Management	Section Total =>				
There is a correct way to do our projects, and a wrong way to do them. We need consistency from all our employees					

We understand the need to define metrics and measure progress, it helps us to get the job done					
We don’t like ‘loose cannons’ in the organization causing all kinds of problems that impact our projects					
When people act independently they mess up the smooth running machinery					
We work best when things remain the same, it’s the needless change that continually messes us up					
Questions	1	2	3	4	5
Integration Management	Section Total =>				
Consensus creates a more accepting atmosphere, and we recognize the value of achieving it					
Without our collaboration we would be unable to complete our work efficiently					
We accomplish much more when we allow people the freedom to collaborate and communicate					

We like to recognize and reward our teams as a group rather than as individuals					
People would prefer to be part of a team rather than stick their neck out alone					
Questions	1	2	3	4	5
Procurement Management	Section Total =>				
Choosing between the project and the people is easy, the people come first					
We develop our people by giving them stretch assignments, it is a necessary cost to the business					
People are more than just a commodity, we focus on them and try to do the right things for them					
If we need to miss a delivery to accommodate an employee, we would be willing to do so					
It just isn't smart to be hardnosed about the project when our people have outside issues to deal with					
Questions	1	2	3	4	5
Quality Management	Section Total =>				

People really listen to different problem solution suggestions					
New ideas are encouraged and embraced					
We recognize and reward risk takers					
It is okay if some of the high risk projects don't succeed					
I would prefer to run a different project rather than run one I know a great deal about					
Questions	1	2	3	4	5
Cost Management	Section Total =>				
People like to act on their own initiative					
We react well to surprises in our projects, especially when we discover something new					
Employees have the authority to act independently					
We find opportunities where other people might find problems					

The way we manage projects is constantly changing and improving when we learn new things					
Questions	1	2	3	4	5
Risk Management	Section Total =>				
We know about our competition and we do everything we can to make our projects more attractive to customers					
We can be ruthless in our desire to complete our projects effectively and efficiently					
We constantly try to understand our customer's needs and adjust our projects accordingly					
When managing our projects, the successful completion of the plan is crucial to us					
Our customers know they can always rely on us to exactly what we promised to do					
Questions	1	2	3	4	5
Scope Management	Section Total =>				
People like to act on their own initiative					

We react well to surprises in our projects, especially when we discover something new					
Employees have the authority to act independently					
We find opportunities where other people might find problems					
The way we manage projects is constantly changing and improving when we learn new things					
Questions	1	2	3	4	5
Time / Schedule Management	Section Total =>				
When people don't follow the rules, we all pay the price					
We have discovered the best way to do things through trial and error, we need to utilize this past experience					
We don't continue to make the same mistakes, because we document the right way to do things and follow the rules					
There is a clear and systematic method for everything, we just need discipline to be able to stay on course					

It is important that people understand the way we do things, otherwise they won't be able to perform well					
Questions	1	2	3	4	5
Teams / Leadership	Section Total =>				
Consensus creates a more accepting atmosphere, and we recognize the value of achieving it					
Without our collaboration we would be unable to complete our work efficiently					
We accomplish much more when we allow people the freedom to collaborate and communicate					
We like to recognize and reward our teams as a group rather than as individuals					
People would prefer to be part of a team rather than stick their neck out alone					
Questions	1	2	3	4	5

Communications	Section Total =>				
Consensus creates a more accepting atmosphere, and we recognize the value of achieving it					
Without our collaboration we would be unable to complete our work efficiently					
We accomplish much more when we allow people the freedom to collaborate and communicate					
We like to recognize and reward our teams as a group rather than as individuals					
People would prefer to be part of a team rather than stick their neck out alone					

Questions	1	2	3	4	5
Human Resources	Section Total =>				
Choosing between the project and the people is easy, the people come first					
We develop our people by giving them stretch assignments, it is a necessary cost to the business					

People are more than just a commodity, we focus on them and try to do the right things for them					
If we need to miss a delivery to accommodate an employee, we would be willing to do so					
It just isn't smart to be hardnosed about the project when our people have outside issues to address					

The Project Culture Map

The project culture map was developed by establishing the dichotomous relationship between the two major factors as shown on the X and Y axes of the map, and considering the continuous nature of the culture that can be described between each of the factor elements. Once it was established that multiple factors would be utilized within each quadrant of the cultural map, descriptors were identified for each of the factor axes. In all cases, the descriptor of the factor was selected in order to establish the focus area of the project team in that region of the quadrant, but it should be remembered that there is significant overlap and crossover of various factors on the map. Secondly, it should be noted that areas of interest appear in the different regions of the cultural map, for example, the use of effective project planning, as each of these focus areas may utilize the planning process in differing ways with regard to that particular factor. As can be seen in Figure 87, the cultural map provides the ability for an individual to graphically describe the culture of the organization as perceived by the survey taker.

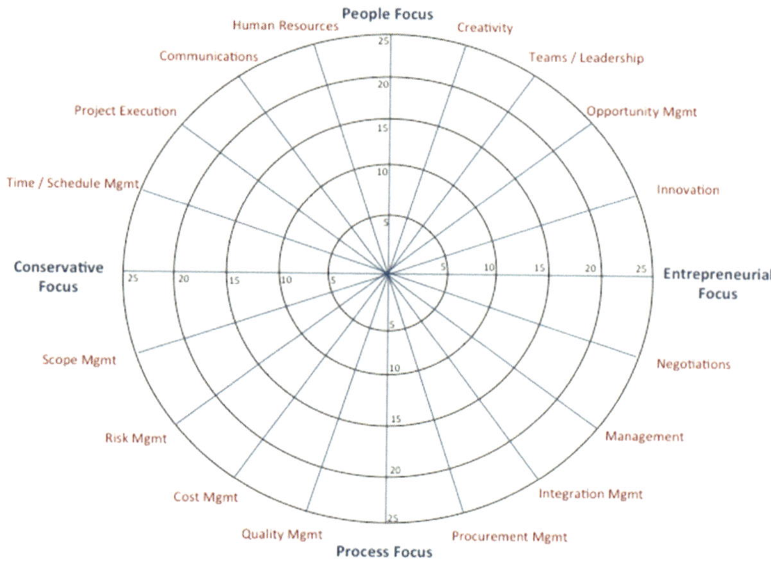

Figure 87: Project Culture Map

Each of the quadrants on the culture map defines very distinct focus areas for the project development team. As an example, in the upper right-hand quadrant, a focus on innovation and technology development is clearly established. This suggests that the companies demonstrating a high degree of importance on technology development would score high within this quadrant area. In the lower right-hand quadrant, the focus of the project team is typically on beating the competition. This competitive focus area will typically drive the majority of actions for project teams scoring high in this area.

In the lower left-hand quadrant, the project teams will be focused primarily on the development of project management processes in order to enhance their performance on projects. Lastly, the upper left-hand quadrant represents the project teams that focus primarily on the well being of the people associated with the project. These focus areas can be seen in Figure 88.

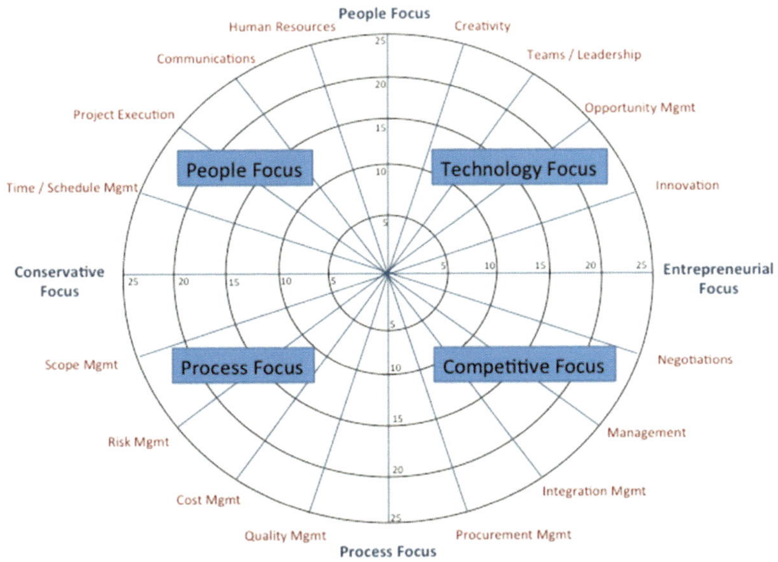

Figure 88: Project Focus Areas

When we consider the general focus areas in each of the quadrants and make a comparison to the nine knowledge areas, we quickly discover that the knowledge areas cover the majority of the left side of this competing values framework, but do not address many of the issues found on the right side of the framework. This leads us to believe that the current model in use by the Project Management Institute is incomplete, and its gaps can be readily identified and understood by means of this proposed project management cultural framework. The nine knowledge areas are identified in Figure 89.

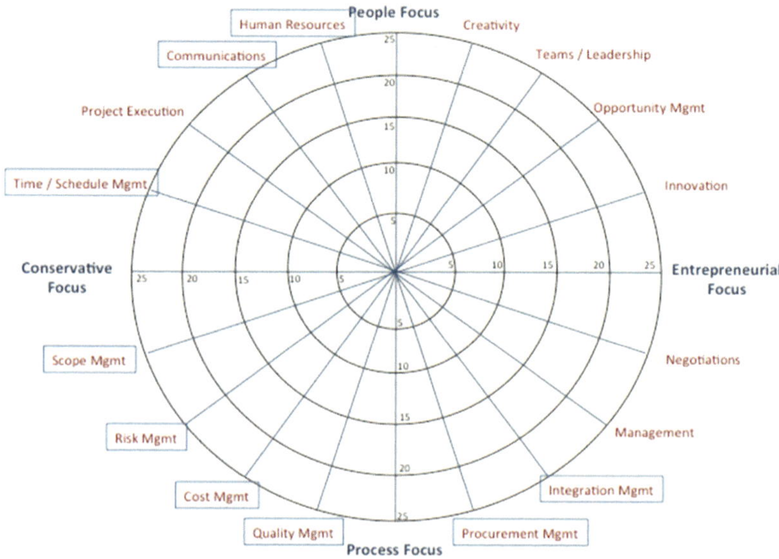

Figure 89: The PMI Knowledge Areas

Interestingly, the nine knowledge areas tend to focus on process and controls with regard to the project. Upon reflection, we realize that even though the human resources and communications knowledge areas purport to deal with human factors, they are limited to processes. Thus, we must identify some of the areas not typically discussed with regard to project management, largely associated with the individual differences and behaviors of individuals on the project team, the group dynamics of those individuals, and the entrepreneurial components associated with the development of new technology and techniques.

Utilizing the aforementioned methods, we have established a new 16-factor framework based upon the independent competing values within a project environment, which will enable us to map the current project management culture for an organization. We will follow an identical procedure as that described for identifying corporate culture, which will enable us to locate and identify specific tasks that can be undertaken to positively impact our project management culture. The process flow is as follows:

Define and graph current project management culture state
Establish preferred project management culture state
Identify cultural factors to be changed
Identify specific tasks to undertake cultural change

The Competing Values Methodology

The competing values framework provides for an ability to assess various factors with a high degree of resolution and then to utilize those independent factors as a taxonomy describing our project management culture. The use of a graphical descriptor enables a rapid visualization of areas of strength or weakness. In addition, the graphical nature of the result allows us to do a rapid comparison between different individuals with regard to the organization's culture. This facilitates a group discussion arriving at a common understanding, and transcends the numbers and details associated with individual factors in the hopes of streamlining the discussion and arriving at a common framework more readily.

To begin this effort, the project management cultural survey is supplied to a large group of individuals familiar with the organization. In most cases, it is expected that this team will be comprised of at least ten individuals, and can be effectively managed in groups of up to one hundred twenty. The steps in the process to generate a common project management culture are as follows:

1. Provide a survey to each individual participating.
2. Each individual completes the survey without discussion within the group.
3. After completing their survey, individuals will graph their corporate culture shapes on the competing values framework as shown in Figure 87.
4. Participating members pair up with another individual and discuss their project management culture assessment shapes with one another.
5. The teams of two will discuss areas of deviation between their project management culture shapes, establishing interpretations and reasoning in order to reach a consensus on a common shape.
6. Once a common shape has been agreed upon, each team will pair with another team and repeat the process.
7. This process is continued until the entire group has produced a common project management culture shape that they believe describes the culture within the organization of interest.

This will generate a shape such as shown in Figure 90, known as the company's current state. Of particular value in this process is the large amount of discussion that will occur between different members of the teams, clarifying interests, understandings, stories and beliefs within the organization. Most companies that have utilized this approach find the large-scale discussions to be of the most significant importance in understanding their project management cultures. Normally, these discussions include a rereading of the questions and discussions involving the reasoning behind the ratings as provided. This effort will hash out any inconsistencies or interpretations that are uncommon within the group. The generalized descriptors for each of the sixteen factors may also be of significant value. Ultimately, the shape is not as important to understanding the project management culture as is the generation of a common understanding that all members of the development team can share, thereby beginning with a common baseline that will enable effective identification of areas to increase on which to focus and improve.

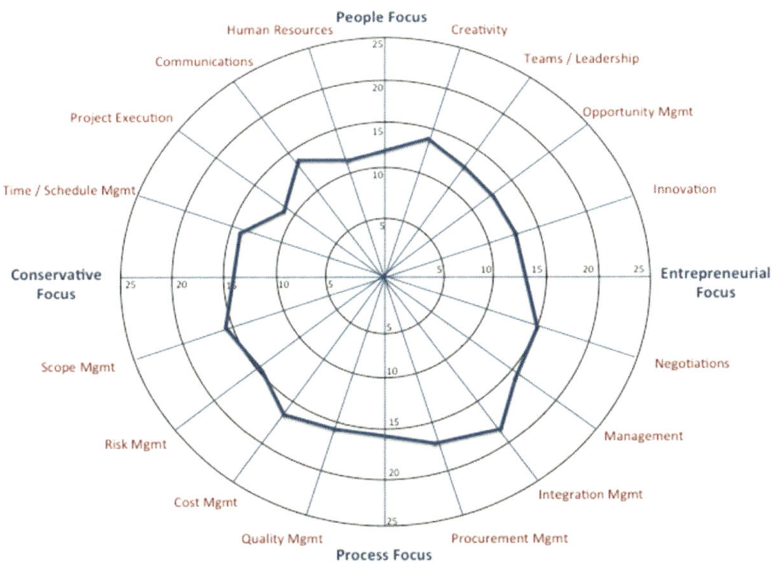

Figure 90: Sample Project Culture Mapping

In the next step, the members participating in this effort must resolve two simple questions:

We have determined the shape of our project management culture; what shape should it be? Why?

This entails a discussion period equivalent to that of the first discussion, establishing a common understanding of what shape would be most appropriate for our organization to accomplish its goals. This is generated over the company's current state, thereby establishing areas that we seek to increase and decrease on the diagram. This will identify areas in which we must define what we are going to start to do in order to increase in that area, as well as areas where we will define what we are going to stop doing in order to decrease in those areas. This will generate a practical action list that will enable the start of project management culture change.

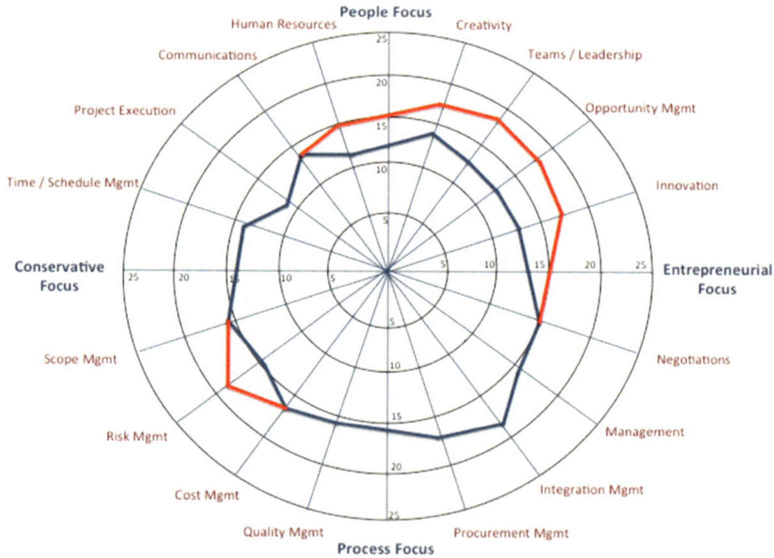

Figure 91: Project Culture Desired State

Figure 91 is a completed cultural assessment and identifies the areas where movement is desired. In this particular case, the organization established six areas of desired change with respect to its current project management culture:

Human resources focus
Creativity focus
Teams/leadership focus
Opportunity management focus
Innovation focus
Risk management focus

Consideration must be given to the amount of effort that is likely to result in beneficial movement with regard to the project management culture. Dependent upon the company, individuals must assess how much change they are likely to implement successfully within the organization. In general, it is best if these areas of interest are limited to three to five factors. However, it is common practice for most organizations to identify far more factors of interest than they are readily capable of undertaking.

Therefore, after deciding which factors will be pursued, individual tasks will be identified as in the previous example. Several considerations are identified for each of the project management culture factors, hopefully providing insight into some potential actions that have proven to be effective on the project management culture.

A Pilot View of the Project Culture Framework

In this section, a cross-section of company types is examined to clarify potential project culture mappings for those company types. An effort has been made to select organizations that have specific characteristics of interest to the reader, and it should be noted that the shapes and descriptions provided are only relevant to the specific companies that have conducted the surveys and may not represent appropriate shapes for similar companies or organizations.

The first type of company investigated is a medical device company, shown in Figure 92. This organization prides itself on an ability to innovate advanced technologies and incorporate those technologies into useful products for introduction into the medical community. The characteristics of the company show significant strength in technology developments as illustrated in the upper right quadrant of the diagram, as well as strength in the lower left quadrant, establishing the need for process and regulatory compliance for those organizations.

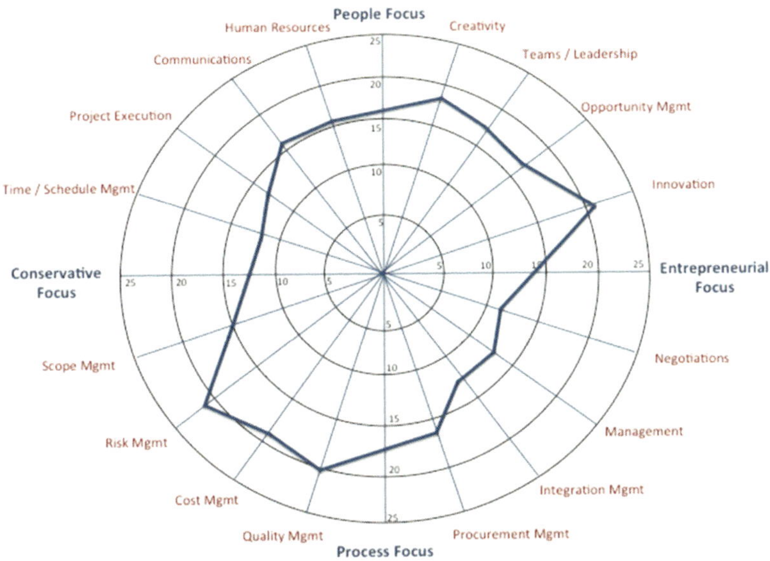

Figure 92: Sample Medical Device Company

As might be expected, there is a significantly lower focus placed on schedule management and the competitive focus that is exhibited by many organizations in saturated industries. Thus, we expect this organization to focus on the introduction of new products on a regular basis, and to spend considerable time with regulatory authorities to ensure that the medical community accepts its products.

The next example, shown in Figure 93, is that of a manufacturing company. The manufacturing enterprise is typically focused upon quality issues in establishing effective processes aimed at reducing both variability and waste. This is illustrated in the importance of the lower left-hand quadrant of the diagram, establishing the need for well-documented processes in the pursuit of manufacturing excellence. Additionally, strengths are noted in procurement management and schedule management, and a significant focus has been placed upon teamwork.

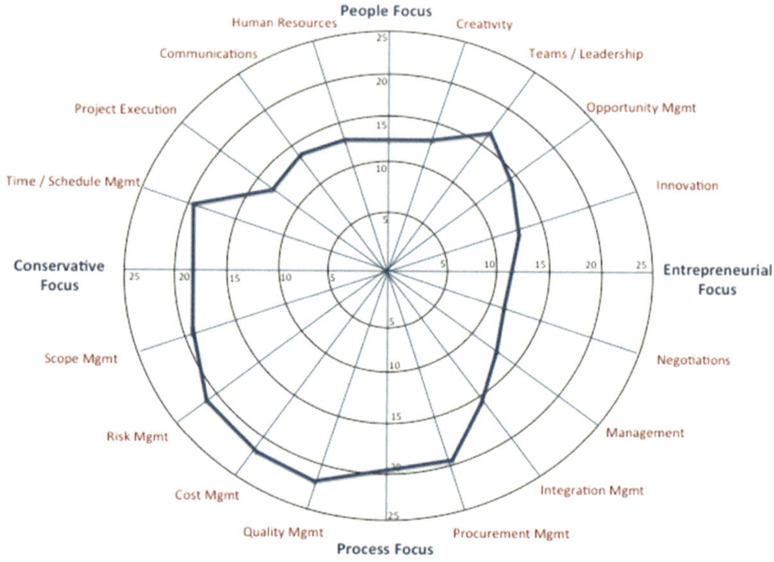

Figure 93: Sample Manufacturing Company

Success in any manufacturing effort is fundamentally associated with the delivery of consistent product with minimal disruptions. In some manufacturing enterprises, there is greater attention paid to the competitive nature of the business, which indicates greater focus in the lower right-hand quadrant of the diagram.

In a service company, shown in Figure 94, a significant focus is shown in the upper left-hand quadrant of the diagram. This demonstrates a focus on accomplishing the needs and desires of the organization being serviced, and is associated with a service excellence business model. While some attention is paid to other areas of its cultural mapping, including a focus on quality and risk management, the principal focus is associated with the people operating on the project team.

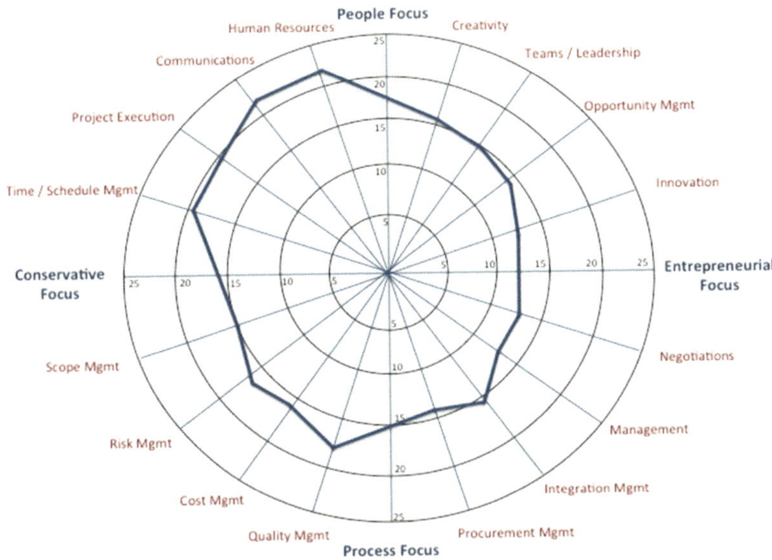

Figure 94: Sample Service Company

This focus aligns well with the customer service model, and will result in a high degree of connectedness between the project team members and the organization being serviced. Relationships in this case are valued more highly then the development of technologies or the need to effectively compete.

A defense company, illustrated by Figure 95, establishes two specific areas of interest. In this type of environment, contracts are awarded by external agencies. By virtue of this, a competitive stance must be taken with regard to the proposal process in an attempt to gain work for the organization. This is demonstrated by the strong focus in the lower right-hand quadrant, where the competitive nature of the organization is illustrated.

The lower left-hand quadrant also shows a significant focus for the defense company, demonstrating the need for accurate documentation and business processes in order to build consistency and credibility with regard to project performance. This is also very typical of a contracting company, which requires that their past performance be considered in the awarding of any new contracts.

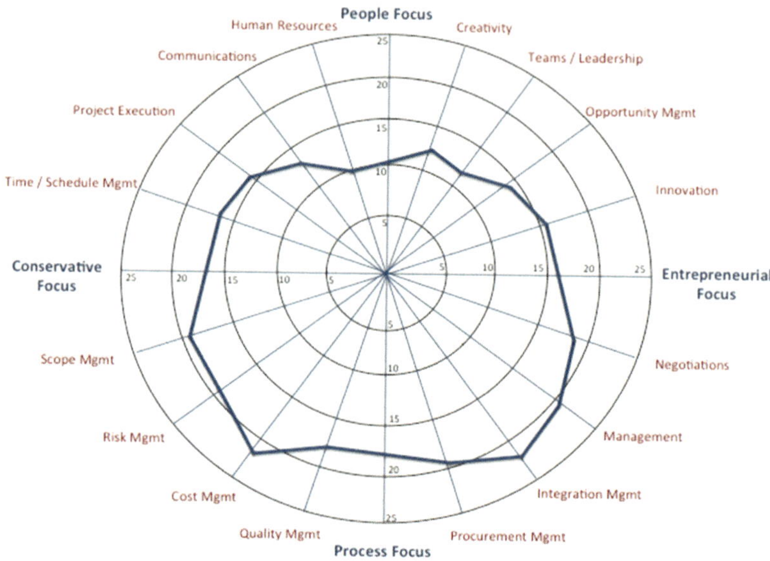

Figure 95: Sample Defense Company

Interestingly, this type of company has a very low focus on human resources. Likewise, limited attention is given to innovative and creative solutions to problems, as these are typically isolated in specific regions of a defense company, such as the research and development areas.

It should also be noted that a larger company might provide a greater diversity of strengths in a cultural mapping exercise such as this. In short, different areas of the organization can demonstrate strengths on this diagram, allowing multiple competitive advantages for the organization.

In a startup organization, as shown in Figure 96, we see the same phenomenon established in the opposing direction. In this case, due to the relatively small size of the start-up company, fewer of the areas register as strong focus areas or strengths. As the organization grows and develops, it is likely that these focus areas will be developed, creating a more balanced organizational culture.

In this particular example, a significant focus is placed on the upper right-hand quadrant, technology development. This should come as no surprise, as most organizations are founded on this strength, and the other focus areas are developed as the organization matures.

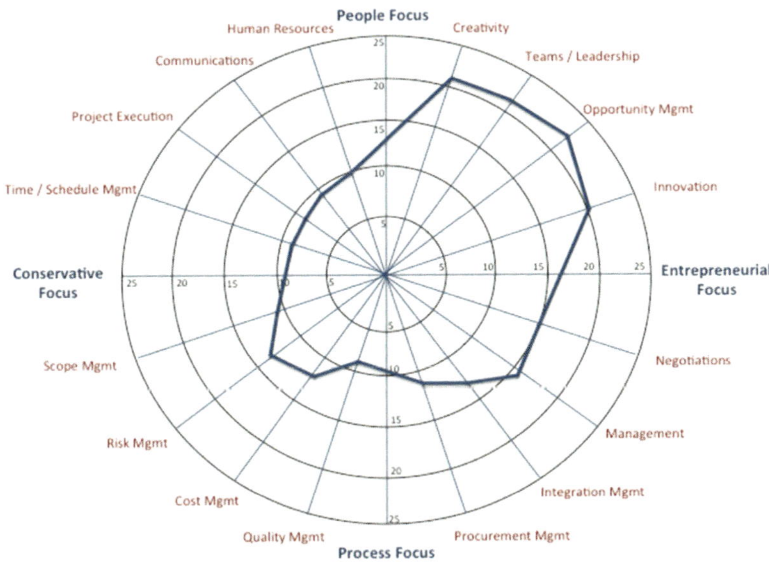

Figure 96: Sample Start-up Company

It is of interest, however, that the use of the cultural mapping technique would enable a start-up organization to identify areas for internal development, thereby maximizing its potential in the marketplace. In this particular case, quality management and schedule management might be appropriate areas for enhancement.

This type of cultural mapping tool may be utilized to gain insights into the challenges faced for project managers. It can also be utilized to help streamline efforts to impact the culture by either enhancing or reducing skill areas in order to better fulfill the organization's goals and strategies. Once again, the most powerful component of this approach is to take full advantage of the rich discussions that will occur during the mapping of the current culture and the establishment of a desired culture mapping.

Considerations for Value Adjustments

Once the gaps of been identified in the project management culture, it is often very challenging for organizations to identify specific actions to be taken that will enhance or reduce the focus in the differing areas of interest. Obviously, there are nearly infinite possibilities that could be utilized with an eye towards adjusting the amount of focus in any of these factors. Ultimately, it is up to the members of the organization to identify the most significant ways in which their project management culture can be reshaped into a more desirable outcome.

A brief discussion follows each of the knowledge areas with some potential considerations that might help to identify areas of improvement for the various project management culture factors. Many resources exist online to assist with these tasks, as well.

Creativity

The study of creativity is full of a variety of resources that enable an individual to improve his or her perspective on creative idea generation. Research suggests that there are a variety of behavioral and situational approaches that can assist in enhancing creative output. It is essential that there are individuals within the organization who are considered to have creative skills prior to undertaking any actions to enhance its creative output.

Place constraints on yourself

In general, people will naturally take the path of least resistance and build off previously developed ideas for existing concepts. Often this leads to lower creativity. By placing constraints on yourself, you can limit your ability to fall back on previously considered notions or approaches. These constraints, whether they are real or imagined, will impact the ability to resolve a problem and will require more divergent thinking to address the issue.

Reformulate the problem

Research suggests that creative people have a tendency to work toward solutions and continuously reformulate problems. Einstein once noted, "If I had an hour to solve a problem, I'd spend fifty-five minutes thinking about the problem and five minutes thinking about solutions." This is also a common method used in a traditional systems engineering approach, as visualization of the problem from various points of view creates much more meaningful insight and understanding.

Keep data gathering separate from ideas generation

In general, people can generate creative ideas with little or no prompting. However, with our hectic schedules and deadlines, we often interrupt our creative processes in order to address a specific element of the problem. This will short-circuit the interconnectedness that can be created by a stream of ideas that build one upon another. It requires discipline on the participant in order to separate the creative process from doing the work.

Maintain a positive viewpoint

Creativity research has discovered that strong positive moods will generate the most creative ideas. This can be particularly effective when generating ideas in a group or project team, as efforts to maintain a positive outlook and an enjoyable atmosphere will often result in richer and more insightful ideas being generated. Secondly, never satisfy yourself with the first or even the tenth solution to a problem. Creativity often requires a sheer volume of ideas to guarantee a rich solution set.

Use free-association

This technique is often used in creativity research, where individuals are instructed to consider an existing system or product, and then visualize its associated problems or opportunities. Free association is largely a right brain activity, not limited by the analytical centers of the brain. Therefore, new potentials and opportunities can be readily identified, some of which may prove to be insightful with regard to developing a breakthrough.

Allow time for percolation

During the development of ideas and approaches to problem solving, we often conduct a great deal of data gathering and collecting of insights with regard to the problem or system being addressed. We must give our brains time to percolate and process this information, usually in a passive activity, in order to gain significant insights and perspectives on the problem. Most intriguing ideas are spontaneously generated in the midst of trying to accomplish something else. Displacing of information within the human brain and allowing it to churn through possibilities and perspectives will often result in significant technological events.

From a project perspective:

- Seek multiple solutions to any challenge
- Ask yourself if the selected solution is the most obvious path
- Accept risk and occasional failure as a component of project management
- Seek a continuous learning environment for your project team
- Recognize that being different is good, and may yield surprising results

Project Execution

Project execution focuses on dealing with the people issues while the project work is being accomplished. This means that an enhanced focus on the individual relationships and group dynamics will yield positive results. There are several techniques recommended to enhance the ability to execute projects.

Spend time with your project team

To gain insight into the true status of your project, the project lead's best strategy is to spend considerable time with the project team. He or she can also spend time reviewing the project documentation the team members have provided. It is likely that this will yield a great deal of information that might otherwise have been overlooked. In addition, the relationships among the team members and their project lead will be enhanced, thereby creating greater insights into the team's perspectives and attitudes about the project. All of this information will help a project lead engage and motivate the project team members.

Use your milestones wisely

It is much more straightforward to manage a project that has been broken down into digestible pieces. This is accomplished by dividing the deliverables into manageable tasks that can be grouped together to accomplish more complex work. As might be expected, increasing the aggregation of the work often increases the amount of insight the project lead can gain with regard to project status and expected outcomes.

Check your assumptions

Project teams often begin a project with careful consideration of the stakeholders' requirements and needs. Unfortunately, those stakeholders cannot often describe what is of greatest benefit to them in the development of the project, except in terms of what they might already possess. For this reason, it is important to remember that many of the decisions and technical approaches selected with regard to the project are based upon a series of assumptions. Those assumptions can be checked and corrected throughout the life cycle of the project, and effective project managers will pay close attention to reviewing and reconsidering previous assumptions with an eye towards enhancing project understanding.

Identify critical skills as well as critical path

An effective project manager will often consider what might happen if one of the key contributors to the project were no longer available. By identifying critical skills with regard to the project, the project leader can begin to cross-train or develop resources other than those identified as critical to provide some amount of risk management with regard to the loss of resources. Most project managers are particularly effective at understanding the critical path of tasks associated with their project schedule, forgetting that critical resources also contribute to the critical path.

Build a compelling story for your project

Lack of engagement of the team, sponsors or investors can often be a critical failure mode for inexperienced project managers. The generation of a compelling story based on the strategy of the organization can yield significant productivity advances and enhance the amount of support provided by external sources. This means careful thought must be provided to consider the essence of the project deliverables as opposed to a specific deliverable to be identified. For example, if McDonald's is focused only on the preparation of fast food, it might be difficult to engage the team members in an improvement effort. However, if we recognize that McDonald's is providing places where families can eat a meal together, we can provide a much more compelling case for our project team, more than likely increasing our focus and productivity.

Don't skip the verification and validation

As project schedules get shortened and difficulties arise during the development phase, it becomes very tempting to short-circuit the testing phases of verification. It is of critical importance to focus on completing all of the work satisfactorily before going live with your project. Therefore, regardless of the pressure, keep in mind that the verification and validation of your work is as critical as the development of requirements were at the project's outset.

Conduct the postmortem on your projects

Much of project management involves guessing and estimating the durations and magnitudes of tasks associated with project development. Effective project managers enhance their abilities over time by paying close attention to their initial estimates for tasks and the actual deliverable times required to complete the work. This self-assessment is often conducted during the project's postmortem, and should be utilized to enhance the project leader's insights and self-awareness.

From a project perspective:

- Provide autonomy for your project team members
- Keep your project team focused on the objectives
- Allow project team members to follow multiple paths to a solution
- Maintain a people-first focus in your project work
- Undertake teambuilding efforts with your project team
- Provide freedom and authority to your project team, as it will often yield significant advantages

Opportunity Management

Opportunities are often problems from a different viewpoint. It is important to understand how to maintain a positive environment within your project team, thereby energizing people's ability to contribute to practical solutions to problems.

Communicate clear objectives

Your project team wants to be part of a compelling future, and need to know what is most important for them to accomplish at any given time. Communicating objectives in a clear and concise manner will allow them to focus on the opportunities that are most important.

Share information with your project team

Keeping your project team apprised of everything that is going on with regard to your project within the organization will often alleviate many of the trivial communication tasks that accompany any project team endeavor. When you keep your team informed, they tend to feel a greater sense of value within the organization. Keep your communications truthful and open, and don't be afraid to share any bad news that comes up. The lack of information often results in a reduction in a team's productivity.

Establish open communication channels

You can understand a great deal about your project team by listening to its members' perspectives and thoughts about the project's needs. Keeping an open mind and encouraging people to feel free to communicate effectively can often be the most beneficial manner in which communications are kept at a high level. For example, a manager should never be offended by a project team member's opinion that his or her latest idea was remarkably stupid. Although this sounds counterintuitive, it is of significant value to find out that you may not quite understand the solution you have proposed; what better time to address that shortcoming than before implementation?

Promote productivity, reputation, and values

Make it clear what on the project is of value to the organization, whether it is highly ethical behavior, high-cost performance, high productivity or many other potential measures. Once your project team members understand the target, they become much more effective at pursuing it in an autonomous matter. Employees are perfectly willing to make significant efforts towards a goal as long as they understand the goal correctly, and feel they will be appropriately rewarded for their efforts.

Build trust

The foundation for any results within your organization rests on a trusting relationship not only between all of the project team's members, but with the project leader, as well. Without trust, it is unlikely that open communications will exist. Without open communications, the project leader cannot hope to fully understand what is transpiring on the project. Therefore, at every opportunity the project leader must seek to enhance the trust inherent within the project team and organization, and in areas where the trust has been damaged, spend considerable effort to repair and remedy that shortcoming. Without trust, a strong team is unlikely to be generated.

Support your team members

You can support your project team members' growth by providing educational and learning opportunities, cross training, coaching and other supportive professional interactions. Always remember that success produces more success. The more frequently you enable your project team members to succeed, the more likely your project will succeed. Remember that it is often the remarkable efforts of your project team that will make the difference between success and failure in your projects.

Delegate properly

Delegation is good for the project leader as it expands his or her ability to accomplish work. Delegation is also good for the employees because it provides them growth and learning opportunities. It also enhances the project leader's ability to communicate with the project team members, as it instills a sense of confidence and builds upon the trust required for the project's success.

From a project perspective:

- Attempt to describe problems as opportunities
- Remain open to new approaches
- Maintain a significant customer focus in any decision-making
- Seek multiple viewpoints on your project
- Maintain a "customer needs" orientation at all times

Innovation

Innovation differs from creativity in that innovation produces real results. In order to enhance the ability to produce real results, researchers have suggested a wide variety of approaches, many of which fall into common classifications.

Iterate

If something isn't working, move on quickly to another thing. The enemy of success is standing still. We must constantly seek to try new and different things in order to generate innovative solutions to a problem. It is often a lack of interest in the switching of ideas that can result in a stifling of the innovation within an organization.

Seek new perspectives

Examining your problem from multiple viewpoints, or asking multiple people to examine your problem can often result in the generation of new ideas. Never be afraid to step back and start with a clean sheet of paper, as it can often yield intriguing and groundbreaking results. As an example, ask the people most familiar with your products and processes to generate a view that removes one of the greatest bottlenecks in their current products or processes. When given the opportunity to free associate potential solutions to the problems, insights can be generated that are not readily available.

Create a sense of urgency

Innovation requires energy, and that energy can be generated by the development of deadlines and the sense of urgency with regard to producing practical results. As an example, as the director of research and development for a large corporation, the author would set deadlines with regard to the required breakthroughs on projects. By setting these practical deadlines, a sense of urgency was created, and in the author's experience, no breakthrough deadline was ever missed. This is a powerful tool for a project leader who has the courage to utilize it.

With regard to project teams, less is more

Do your best to identify only those people most critical to the project's success when assigning your team members. The development of a viable and effective project team is not intended to be an effort at empire building. Instead, if you keep your team small and focused, you can avoid many of the conflicts and consternations typically exhibited by large groups of people working together. Of course, due to the complexity of your project, the team might need to be quite large. The project manager must determine the most appropriate size for the team, keeping in mind that when dealing with people, less is more.

Welcome ideas

There are many human behaviors and corporate behaviors that get in the way of accepting new ideas. You must pay very close attention when people are critical of other individuals' ideas, and to the language that is chosen with regard to team meetings. Always focus on disassociating the ideas from the personnel, and on removing critical comments regarding the generation of new ideas. Both are remarkably detrimental to the generation of innovative solutions to problems.

Hire well

It is always better to have one great employee than three mediocre employees. Seek out individuals who have demonstrated a need to innovate in their past experiences or in their desired occupations. These are usually individuals who are comfortable with ambiguity and are willing to accept the risk of trying new things. Innovation is a rewarding experience built on a mountain of frustrations.

Trust your team

Project leaders must have the courage to allow their team to go down multiple paths, and to allow them to try crazy things in their attempts to meet innovation goals. This requires openness and trust from the leaders, and willingness to be courageous in defending innovations when they don't turn out to be ideal. It involves a lot of interactions with the staff to demonstrate that you are invested in them as technical leaders of the organization. Keep in mind that virtually everything your organization sells today was an innovation sometime in the past; we need to keep the pipeline full and the ideas percolating.

From a project perspective:

- Require more decisive behaviors
- Accept failures as a cost of doing business
- Seek autonomy for your project team members
- Recognize that innovation is different from creation
- Embrace change as a desirable action
- Recognize that differences in perspective may yield significant returns

Negotiations

Negotiations are crucial to the success of any project effort. This is due to the challenges faced by every project manager; managing contradictory requirements and dealing with differing viewpoints among the project's stakeholders. An ability to effectively negotiate such troubled waters will result in a highly successful project with satisfied stakeholders. There are several items of interest to any negotiator, regardless of situation.

Make sure you know what the stakeholder is looking for

Having insight into the needs and desires of anyone with whom you are negotiating is crucial to a successful transaction. We often classify the other party's interests with regard to our own, and have a tendency to see the other party as an enemy. This is counterproductive; as upon reflection, we may realize that we are willing to give the other party what he or she wants. By spending time reflecting on the other party's wants, we better understand how to formulate and execute an effective negotiation.

Keep your head in the game; don't get sidetracked by the other party

We have many emotional responses that we must address in a typical negotiation. Often these responses are governed by judgmental heuristics that might limit our ability to rationally consider various options and perspectives. By understanding how these judgmental heuristics can get in the way of our ability to make effective decisions, we can better manage the situation and more capably control the information that we utilize. Therefore, we should always attempt to be thoughtful in our approaches in negotiations, slowing down the transaction to enable our brain to percolate the possibilities. This has an additional beneficial effect, as well, as the other party might become uncomfortable with the slowing of the negotiation, thereby providing some advantage to the individual controlling the pace.

Find alternative solutions that benefit the other party

We often enter a negotiation with the viewpoint that the other party is the enemy. We do ourselves a significant disservice when we make this basic assumption at the outset of any transaction. Likewise, the other party may be making a similar assumption about us. If, however, we spend some time in the generation of alternatives that will benefit the other party as well as ourselves, we can defuse this potential problem before it becomes an issue within the negotiation. By demonstrating that you have the other party's interest at heart, you demonstrate that you can be trusted and are likely to seek a solution that will be mutually beneficial for all involved. This increases trust between the parties involved in the transaction and will ultimately result in a more optimized solution.

Understand what success looks like for you as well as for the other party

We must take the time to understand fundamentally what we seek out of the negotiation prior to undertaking the transaction. While this may sound ridiculous, many negotiators spend little or no time in preparation prior to the negotiation. By virtue of this, every potential solution sounds attractive. If we consider what success looks like for the other party, however, we can better shape our approaches to maximize the potential for acceptance by both sides. In addition to this, we will generate trust with the other party and will more readily produce acceptable solutions for both parties.

Don't be afraid to use silence

It is a significant shortcoming among negotiators that they do not understand how to use silence. When faced with a challenging situation, silence can be used to control the pace of the transaction, as well as to limit the brain's ability to react emotionally to the transpiring events. Commonly, when an individual employs silence during a transaction, the other party jumps in to fill the void, often making concessions or providing information that would not otherwise have been available.

Never put anything in writing until you are ready to sign the contract

When negotiating, always keep your head in the game. This implies that you maintain your ability to think rationally regardless of the emotion or tenseness of the situation during the transaction. This also means taking the time for thoughtful reflection and managing the communications in a manner that will enable you to establish your points and clarify your understanding of any other party's perspective. Rather than record a partial solution in writing, it is best to continue working through the transaction until both parties are close to agreement. At this point, recording and documenting the transaction will be of significant benefit, whereas anything recorded before this time may be subject to change.

From a project perspective:

- Approach conflict resolution as a critical factor for your team
- Generate a communications plan
- Conduct your project with a distinct business rhythm

- Focus your negotiations on mutual benefit
- Reward collaboration and cooperation on your project team
- Seek openness in solutions generation

Management

There is a significant difference between management and leadership. Managers have a specific focus on utilizing appropriate processes and approaches in accomplishing the work that must be completed, while leaders often do not follow a specific process but are capable of accomplishing great things. There are many items of interest to us as managers that will cross these boundaries, and are critical aspects of our project management success.

Define your vision and communicate it

Many years ago when you came to work you were expected to leave your intellect at the door. Managers did the majority of thinking for the organization, and the employees were simply there to accomplish work. In today's knowledge environment, it is important to engage project teams to get the best of their productivity and abilities. This requires a clear vision of your organizational strategy, as well as of your expectations regarding the project's outcomes. Take time to develop an effective vision and communicate it in a concise and understandable way.

Get to know your people

Communicating downward is an excellent way to establish expectations, but it is not enough to engage the team and gain their commitment. Getting to know your project team will pay large dividends in the future. It is the responsibility of the project lead to establish relationships and maintain them in a manner that will enable the project leader to fundamentally understand the needs and desires of all of the project's contributors. Having a personal connection with the individuals performing on the project team goes a long way towards gaining the commitments and trust of the contributors. Take time to know your people, and you will save time in the long run.

Develop your ability to influence others

Leadership is often described as the amount of credibility that you have with your project team. Often, this is created through apocryphal information and stories. Take the time to pay close attention to the stories that you choose to share with your project team, establishing for them the importance of execution on the project team. Keep in mind that your goal as a leader is to find people who are willing to stop doing what they're doing and follow you. This requires a change in direction and a positive attitude on the part of both the leader and follower.

Develop excellent communication skills

The secret to great communications is to communicate concisely and frequently. This means creating a culture in which your employees can share information freely and work with you to accomplish objectives. This will increase their engagement on the project team and the probability of successful completion of your objectives. Ask them questions and respond positively when they ask you questions.

Utilize the power of gratitude

If you spend a moment considering past managers who have earned your respect, they're likely to share a significant trait: gratitude. Saying "thank you" or "well done" to an employee who has successfully completed a task is a very powerful motivator, and is ranked higher than money on most employee satisfaction surveys. Spend time practicing this skill until your positive feedback becomes a habit.

Focus on job satisfaction

Many companies have a long history of losing their most talented employees. The cost of recruiting talented employees is extremely high. Therefore, it is crucial for managers to focus on employee job satisfaction. Typically, this is accomplished by regular communication, but in the event that there are issues arising, it is incumbent on managers to recognize the power they have to increase their employees' job satisfaction:

- Maintain a respectful environment
- Express your gratitude
- Make sure you are clear with your expectations
- Provide enjoyable work
- Allow flexibility in scheduling
- Make sure your employees are rewarded as they deserve

From a project perspective:

- Establish a well-documented project management process
- Maintain consistency in your behaviors towards projects

- Establish effective metrics to manager projects
- Demonstrate the importance of keeping the project on plan
- Share effective strategies for project success

Integration Management

The focus of integration management on any project effort is intended to establish a seamless introduction of the project objectives into the organization's existing systems and products. This involves several crucial factors, and it is the responsibility of the project manager to lead the activities to provide a framework for individual contributors on the project.

Develop a coherent project plan

The project manager must take adequate time at the outset of the project to focus on the needs and requirements of all of the stakeholders. These must be assessed and analyzed in order to produce a requirements list that will be utilized for the project's execution. When the requirements list is put together, traceability must also be maintained to ascertain where the requirements were sourced. Upon completion of the requirements, a detailed specification is defined, and from that flows a work breakdown structure and schedule. Risk must be applied to the schedule to increase the accuracy of the project estimates, and finally a budget will be extracted from the schedule and resource assignment information. These documents are critically important in communicating the project's expectations to its team members, and they allow the project leader to maintain an effective status at any given time during the execution phase.

Execute the project on plan

Once the plan has been completed, it is the project manager's responsibility to ensure that all members of the project team are conducting themselves in a productive and effective manner with regard to the outlined project plan. This is typically accomplished with significant communications and regularity of meetings, maintaining status, and ensuring that any issues that require attention are immediately addressed. The project execution must always be accomplished with an eye towards the strategy of the organization, ensuring alignment of the project objectives with those of the organization.

Manage changes throughout the life of the project

Once a project plan has been developed and communicated to the project team, the project leader must understand that it is only a general framework for the project's completion. Due to enhancements in knowledge and perspective throughout the life cycle of the project, it is likely that many of the items established within the project plan will require change throughout the execution phase. The project manager must not only embrace this change, but must manage the change effectively within the project documentation to ensure that coherence on the project is maintained. This is accomplished by holding frequent meetings to review the plan, and through the use of an effective project documentation system.

From the project perspective:

- Establish a well-documented planning process
- Establish templates for planning documents

- Manage a stringent process control on your projects
- Establish detailed schedules with resource allocations early in the project
- Maintain a common understanding of the project team by frequent reference to the project plan

Procurement Management

Buying products and services for your project is an important discipline to develop. Failure to accomplish this effectively may result in unnecessary costs and challenges.

Manage procurement with invoices

Implementation of an invoice schedule is important for an understanding of how expenses will be incurred on your project. When invoices are received on a recurring schedule it is much easier to monitor them and ensure that they are paid in a timely manner. Take care to validate your invoices' prices and any changes associated with the pricing, as inspecting the bottom-line cost is not adequate to ensure that you are managing your project with good financial responsibility. An invoice database is often utilized in order to forecast project expenses and track unusual expenses. This is easily accomplished with a simple spreadsheet.

Manage your contracts

When conducting complex work, it is common for some of the work or products to be outsourced via a contract. Establish early in the conduct of the business that you will provide significant oversight to any contractor. This oversight might be as simple as regular meetings and status reports from the contractor, and may extend well beyond this to regular audits and assessments of the work being accomplished. Make certain to apply all organizational standards and processes to the subcontractors through the contractual documents. This will ensure that the standards and policies are followed by the contractor, and consistent product will be delivered that will require little or no integration on your project.

Manage your bids

A poorly executed bid process creates a great deal of consternation within your organization. Take care to run a consistent and fair process to avoid any concerns about trust or wrongdoing. First, review the pertinent requirements of your organization and ensure that the contractor can fulfill them. Second, document your entire buying decision process to provide traceability in the event someone challenges your selection in the future. This allows you to defend any decisions made with regard to suppliers selected or declined.

From the project perspective:

- Maintain openness with all suppliers
- Remain open to new vendors at all times
- Establish vendor performance requirements
- Establish partnering documentation
- Develop a systematic selection process

Quality Management

Quality is the name of the game in the project management world. This is what separates the professional project manager from the amateur. The commitment to produce on time is admirable, but it is not enough to ensure that project objectives are met appropriately. A commitment to quality will make a big difference in your project management efforts.

Commit to quality

An organization's commitment to quality has to come from the top, in this case, the project manager. It needs to be reinforced on a regular basis and established as a non-negotiable goal of the project. Quality decisions should not be situational, and project leads should never accept inferior quality in the work being accomplished on their projects.

Track deviations

The level of quality expected must be established early. This may include a measurement of deviations from schedule or budget, or the quality of products being produced through the execution phase of the project. Therefore, the project lead must set standards early and regularly communicate performance with regard to standards.

Train your staff

Institute a new employee initiation program that will establish for all project team members the need for focusing on quality issues on a daily basis in the execution of their jobs. In general, it is considered good practice to put existing project team members in charge of training all new team members, because they can provide a first-hand perspective on how projects should be executed properly. Quality is not to be managed only by the quality department, but by every individual contributing to the project.

Establish quality improvement teams

Organizing project team members into quality improvement teams can be a very effective way to identify and address problems before they occur. These are typically groups of employees who are encouraged to assess existing processes and make recommendations for improvement, all with the intention of improving the output of the project's deliverables. This requires that the team be given both the authority and responsibility for making the project better. It is incumbent on the project manager to provide the team with adequate resources to establish their analysis and the ability to effectively communicate their findings to the project team.

From the project perspective:

- Establish quality expectations early in the project life cycle
- Focus employee efforts on the reduction of waste and very ability
- Revisit the voice of the customer on a regular basis in project meetings
- Establish the nobility of reliable performance of the project team
- Manage changes to the project plan effectively and on a timely basis

Cost Management

Establishing rigorous cost management processes early in the development of the project is crucial for enabling appropriate status information and insight into the likely outcome of the project. This is normally accomplished by following a simple process:

Establish your current state

You cannot understand the where your project is going or its likely outcomes unless you can effectively communicate its current state. Gaining insight into what you're currently spending to accomplish deliverables is an important first step towards properly managing those costs.

Identify inefficiencies

Understanding and highlighting a few areas where you can make improvements will help to focus your efforts and create the greatest return for the investment in any improvement program. This is often difficult to accomplish if you are busy putting out fires every day, and so it should be incorporated as a regular component of your project team meeting.

Manage project cost with standardization

Many organizations feel that their initial view of project budget will likely be realized at the completion of the project objectives. Project managers, however, understand that initial projections are unlikely to be reflective of actual occurrences. Due to scope creep and ineffective planning or inaccurate estimates, you may find yourself challenged with pressures on the budget. Therefore, it is important to standardize the process to continually revisit the project's costs to manage deviations from expectations in a timely manner.

From the project perspective:

- Establish and earned value management system on your project
- Track timecards effectively on your project team
- Establish a business rhythm to examine the budget on a regular basis in project team meetings
- Maintain your project team's focus on the bottom line
- Utilize current performance as a predictor of future performance on your projects

Risk Management

Risk management provides the project manager with insight to the potential liabilities facing the project with regard to uncertain futures. In fact, the majority of effort on the part of the project manager during the execution phase will be related to managing the risk register.

Ask the right questions

High-performing project managers will frequently ask probing questions about uncertainties within the project's deliverables. These must be focused on all stakeholders' perspectives, and should extend well beyond the interests or perspectives of the project team. Keeping a project in control is a misleading concept in a largely unpredictable world. When presenting strategies and project plans, everyone in the organization should recognize that the future is inherently uncertain and that the infinite possibilities are too complex for anyone to predict with great accuracy. With that said, the project manager must make predictions to the best of his or ability, but we all need to be practical about our expectations of those predictions.

Focus on culture

Many organizations have allowed the development of toxic behaviors within their walls. This is an insidious force that tends to sub optimize the performance of the organization. In general, organizations will benefit from clarity on what is expected from all members of your team. Your communication regarding what constitutes acceptable behavior, as well as acceptable deviations from the stated objectives, must be provided. The project team should conduct discussions about the level of control it is practical to manage on the project, and the expectations of stakeholders throughout the execution phase. Care should be taken to manage those expectations to reasonable levels.

Use appropriate reward systems

Reward systems must be focused on the behaviors of interest for the organization. This requires that managers understand the behaviors of interest with regard to their employees, and that they take the time to help steer their behaviors in desired directions. Project managers must lead by example and ensure that fairness is present in all of their transactions within the organization. Doing so will encourage the team members and increase the amount of trust generated on the project.

Focus on the organizational strategy

The primary purpose of all risk management is to contribute to the realization of the organization's objectives. Project managers need to communicate that these objectives are developed in order to create and preserve value for the stakeholders in the project. These stakeholders are the providers of both project objectives and the financial support necessary to continue development efforts in the organization. It will be of benefit to any project messenger to take the time to communicate this alignment during regular project team meetings.

Understand your limitations

Risk assessments are simply your opinions about the future. Given adequate aggregation, these analyses will often yield predictable results with regard to the organization, but one should never forget that they are simply guesses based upon the estimates, which are based upon uncertainties. To determine the extent to which the organization is ready to deal with potential future events, risk analysis needs to include matters that can aid the realization of the business objectives as well as those that might cause negative impacts. Project managers should recognize that these analyses are nothing more than indicators of potential paths in the future, and must be managed on an ongoing basis throughout the project execution phase.

Maintain consistency across multiple projects

Project managers should establish straightforward integrated reports, and require their execution on all projects within their responsibility area. The aim of these reports is to build a shared view of the extent to which the business objectives are at risk and potential corrective actions that might be undertaken. Project managers should also monitor the effectiveness of the risk management system, ensuring that estimates become more accurate and predictive as the project execution progresses.

From the project perspective:

- Understand that uncertainty is a necessary component of any project
- Utilize risk analysis in order to estimate the potential financial liability of the project
- Establish a risk register
- Manage the risk register on a weekly basis
- Establish with the project team that their efforts are to draw down risk during project execution

Scope Management

Project scope management is often the most challenging aspect for any project manager. Everything delivered by the project is defined as the project scope. This lays out the baseline of accomplishment, but any effective project manager also understands that scope will continue to change throughout the life cycle of the project. Effective scope management is not intended to minimize those changes, but rather to incorporate them seamlessly into the project plan as the project progresses.

Spend time documenting requirements

It is difficult to spend between ten and twenty percent of the duration of the project budget meeting with stakeholders to understand the project's requirements. However, projects whose managers fail to do so are often considered failures in the final review, as requirements are necessary to keep everyone working coherently. These must be documented clearly and must clarify what will be accomplished. This document must be shared with all members of the project team and with the stakeholders, and becomes part of the project charter. As changes are incorporated, the project manager must republish this document on a regular basis.

Prioritize your requirements

An exhaustive list of requirements can be a daunting subject for many members of the project team. It therefore becomes important to prioritize them to ensure an understanding of whom the requirement addresses and why it is needed. In the event that some requirements cannot be met, this allows the assessment of impact to the acceptance of the project based on missing them.

Define your milestones with regard to requirements

Many project managers do not draw a line between the deliverables and the milestones associated with the project plan. In fact, the delivery of one of the objectives of the project is largely what defines each milestone. In order to reduce the complexity of very involved projects, many project managers will define smaller milestones and intermediate goals in order to keep their project on track. Typically, with greater aggregation comes better control for the project manager.

Maintain regular communications

This is a key aspect of scope management. As modifications are incorporated into the project plan, they must be communicated to all stakeholders. Keeping in constant communication with the stakeholders and the project team ensures that everyone understands the current requirements of the project and remains aligned. Remember that surprises are typically not well accepted at the conclusion of projects.

Manage changes

All project managers understand the necessity of scope creep throughout the life cycle of the project. The best that any project manager can do is to be prepared. Define an effective change management process early in the project and ensure that your team adheres to it. In addition, take time to forecast changes that might occur in the near future, and ensure that change management is conducted in a coherent way with the existing project plan.

From the project perspective:

- Establish detailed planning documents early in the project
- Establish a communication plan
- Establish documentation standards for project artifacts
- Establish a change management process to ensure coherence inconsistency on the project
- Communicate the need for adaptability and flexibility to your project team.

Time/Schedule Management

For most project managers, schedule management is the name of the game. While we seek to manage our scope, schedule and budget, in some organizations scope can be a very nebulous item. In other organizations, budget may be of less importance than meeting the scheduled end date for the project. Therefore, most project managers seek to focus on schedule adherence as best they can. There are several factors we can keep in mind in order to improve our schedule performance.

Establish your schedule status

The project manager must take time to understand the current status of the project, tracking it against both the cost performance and the schedule performance. This enables the project manager to make predictions with regard to the likely outcome of the project, and helps him or her to understand the magnitude of deviations from the project plan.

Identify reasons for deviations

Part of the project team meetings should be dedicated to identifying and analyzing the reasons why the project team is not adhering to the schedule. They might be related to unrealistic estimates, absenteeism, turnover of employees, or work not being accomplished at an appropriate rate. Once the reasons for deviations have been identified, the project team can brainstorm to devise methods by which they might resolve the challenges and improve the overall project performance. It is important for the project manager to establish the expectation that schedules will be met during the execution of the project.

Communicate why the schedule is important

Project team members might not understand the business impact of not meeting the schedule. The project manager should explain the cascading effect of a late schedule in any complex organization. Often, the inherent costs of the cascade exceed the value of the overall project. For this reason, it is expected that project team members will take the organization's promises seriously and make every attempt to adhere to the project schedule.

Conduct schedule reviews on a regular basis

Project team members are more likely to pay attention to items that are brought to their attention on a regular basis. This is one of the reasons why schedule performance and the risk register should be revisited frequently during project team meetings. The project lead must take the time to identify deviations from schedule plan and their causes, and communicate forward the actions to be taken with regard to correction of the deviations or correction to the project plan.

Be realistic

The project manager must be very realistic with regard to expectations on project performance. It is a common behavior of inexperienced managers to inform their staff that they are expected to achieve any arbitrary schedule, with little explanation given for the methods by which that might be accomplished. This very quickly creates an atmosphere of mistrust between the project lead and the project team members, and should not be tolerated. Project managers must be aware of the stringent nature of their schedules and consider appropriate behaviors of their teams in performing to those schedules. Without this consideration of the project team, it is unlikely that productivity will be maximized on the project.

From the project perspective:

- Establish a detailed schedule with interdependencies and resource requirements
- Establish the importance of keeping promises on the project
- Utilize the schedule status to make predictions with regard to future performance
- Continuously track the status of the project
- Communicate the status on a regular basis and provide feedback on progress achieved

Teams / Leadership

Effective project teamwork is an essential component for any complex endeavor. However, teamwork and leadership are often at opposite ends of the spectrum. While acting as a project leader, it can be difficult to facilitate consensus building and team building activities. Therefore, we must take time to focus on the development of a team atmosphere for our projects.

Conduct team training

Project managers must understand the value of individual contributors and their unique contributions to the project's deliverables. Some time should be taken to explain to all project team members the importance of each and every other member of the team, and their diverse skills and viewpoints.

Manage conflicts fairly

As a project leader, you will be responsible for managing conflicts as they arise on the project. Take time to work with the team to understand the value of working towards common objectives while seeking fairness in transactions involving disputes amongst project team members. This allows the building of relationships between team members and increases the level of trust and loyalty they share.

Facilitate mediations

As a project lead, remain open to suggestions and concerns brought forward by the project team members, with a particular awareness of the emotions of the individuals. Communicate as clearly as possible, maintaining clear expectations and delegation.

Define the team operating rules

Early in the project, take time to establish the expected behaviors for each of the team members. Ensure that team members understand when and where meetings will occur and what information is to be provided by them at the meetings. Communicate as concisely as possible, and require that your team members do the same.

Establish leadership expectations

While it might be interesting for each and every team member to have a say in all decisions made on the project, project managers understand the need to reserve the right to override any decision defined by the team. It is important to minimize the need for deviations from team recommendations, but the leader should quickly establish the responsibility in the event this must be accomplished.

From the project perspective:

- Establish a method of consensus for your project team
- Communicate the importance of collaboration on your project
- Focus on employee engagement with regard to an effective project plan
- Provide your project team with autonomy and authority to ensure their engagement
- Seek methods of achieving synergy on your project team
- Pay close attention to group dynamics in any transaction

Communications

Communications are a constant challenge in any project environment, largely because of the diverse perspectives of the project teams. It is essential to improve communications as far as possible to streamline the team members' effectiveness in accomplishing their objectives.

Have one conversation at a time

Many people are under the impression that they are able to multitask. However, instead of accomplishing multiple tasks simultaneously, they are splitting their time and dividing their attention. Very rarely will a project team member approach you and ask for your divided attention. We must be respectful and aware of each and every individual seeking to communicate with us to build trust and loyalty on the project team.

Listen carefully

Communications are designed for the purpose of transferring information and gaining insight with regard to the project. Unfortunately, no method of communications is entirely perfect, and it is important to ascertain if the other party understands the intent of your original message. Communications can be improved by listening very carefully and asking open-ended questions to seek understanding on both sides.

Write it down

Although it may be tempting to try to commit everything to memory, it is far more useful and reasonable to record items in writing. Limitations in the human brain and imperfect recall of events reduce our capacity to maintain information in an unbiased manner. However, if information is immediately recorded and dated in a document that can be retrieved at a later time, misconceptions and miscommunications can be reduced or eliminated within the organization.

Establish a communications plan

Setting up a business rhythm for routine communications on your project can improve the organizational productivity and will help with managing expectations of the people on your project team. Spend time communicating expectations, as well as times when responses are required. This ensures that people are kept informed on a regular basis, and defines that the project manager expects similar behaviors from the project team.

Always assume the best intentions

With increasing technology options, it is often difficult to understand the true intentions of communications. It is always in our best interest to assume that the sending party had the best intentions when they initiate contact, as this will often reduce consternation with regard to communications that might appear to be offensive or inflammatory. In the event that you are unclear on the intent of the communication, don't be afraid to resort to more direct forms, such as a telephone or face-to-face meeting.

From the project perspective:

- Establish a communications plan early in the project
- Establish a business rhythm with regard to communications
- Conduct regular project status updates
- Conduct training to improve listening skills
- Pay attention to body language and tone
- Seek opportunities to improve connections between team members

Human Resources

Your human resources are the greatest assets you possess on the project team. The job of the project manager is to maximize the value of those resources in order to produce the deliverables of the project with the lowest investment possible. This must be accomplished while maintaining the respect of the project team and instilling a sense of trust and loyalty in each participant.

Be aware of all of your assets

Often, project managers focus on obtaining resources who exhibit specific skills or traits desirable in the execution of the work. We all would select the stars, given the opportunity. However, project managers can easily forget that they have a number of individuals assigned to their project that may require development and training. An effective project manager makes every attempt to utilize all of the skills of the assets at their disposal. Sometimes, challenges are manifest by the project manager in deciding that certain members of their project team are not desirable or attractive to have assigned to their projects. It is almost always possible to approach these less valued associates and enhance their skills and performance through careful consideration of their needs. This is an essential responsibility for every project manager.

Remember that fairness means treating everyone differently

Project managers must remember that every project team member is an individual with his or her own specific and unique differences. It is, therefore, important to take a personalized approach to the day-to-day management of any project team member. Understanding and taking into account your staff's personalities and preferences when managing and communicating with them yields significant benefits, trust and loyalty. By managing these factors appropriately, productivity on the project can be maximized.

Take your communication seriously

Every project will have to deal with changes and interruptions that will directly impact its team members. Project leaders must manage communications across the entire project team and with stakeholders to ensure that disruptions of this type are public knowledge and can be appreciated by the various members of the project team. This will go a long way towards reducing conflicts amongst project team members due to misunderstandings with regard to expectations and deadlines. Moving quickly in these sensitive areas yields significant results for the project leader.

Recognize that change is imminent

No matter how hard we try, changes will be incorporated on a regular basis to any dynamic project. By definition, these will cause disruptions and challenges for the project team members, and may cause consternation and conflict within the project team itself. Successfully managing and communicating change is an important skill developed by an effective project manager. Seek to minimize misunderstandings and misapprehensions by addressing them head-on.

From the project perspective:

- At all times maintain a focus on your project team members
- Build relationships with your project team members in order to understand their needs and desires
- Develop people as they work on your project, as it will result in significant returns

- Maintain fairness in all activities associated with the project
- Communicate the importance of being sensitive to other project team members
- Establish a change management process to ensure the project team remains adaptable and flexible

Undergoing Project Culture Change

Organizational culture governs virtually all aspects of what is accomplished in any given organization at any given time. Other companies can copy our strategy or methods, but culture is something that arises out of the self-organization of the system, and is inherently embedded in the system. By virtue of this, culture is an extremely difficult entity to change, and when changes are undertaken, there are often many correcting forces that will be activated within the organization itself.

Culture has been identified by a wide variety of descriptors, Including:

Culture is how we do things
Culture is the collective values and behaviors that contribute to the unique social and psychological environment of an organization
Culture is the sum of values and rituals that serve as the glue of the organization
Culture is a system of shared assumptions, values, and beliefs that govern how people behave
Culture is civilization in the workplace
Culture refers to the philosophies, attitudes, policies, behaviors and practices that define our organization

Fundamentally, culture is created when a group of individuals coalesce in order to accomplish an outlandish goal. This might be the founding of a company or the resolution of a particularly challenging problem. In any of these cases, individuals must come together with their collective understandings, knowledge, educations, belief systems, attitudes and perspectives, and somehow produce a meaningful and coherent whole from their disparate group of individuality.

Chaos theory tells us that organizations made up of a large number of independent agents will spontaneously coalesce into an organization. It will be governed by a variety of rules, procedures, policies and perspectives, all of which make up the foundation of the organizational culture. In short, we got the way we are because that is how we are.

As the reader might expect, culture is inherent in everything we do, based upon past experience and methods that have been developed by those who have gone before us. Because of this self-organization phenomenon, the reason why the specific rules and methods have evolved in the way that they have is not always clear to the organization's members. The making of the existing organizational culture is what happens every day. Without an understanding of why the culture has formed in the way it has, it can be difficult to understand how to embark on cultural change.

Fortunately, we have discussed the use of a tool that allows us to utilize a taxonomy of the project management culture in our organization, identifying specific areas of interest and ultimately creating actionable tasks to be utilized to embark upon an organizational cultural revolution.

Frequently, however, this is not adequate to implement practical cultural change. We must also understand that there is significant impetus built into the organization and correcting factors will frequently drive us back onto the original path as we attempt to make changes to culture.

Therefore, we must maintain a laser-like focus on the tasks we choose to implement, insisting that the way that we have done things in the past is not necessarily the way we choose to do them in the future. This implies that we must become change agents within the organization, embracing the fact that differences are desirable in our behaviors, and with the guidance of a cultural mapping tool we can focus our efforts in the areas of greatest return.

With the understanding that cultural change will be perhaps the greatest leadership challenge we will ever face in our professional careers, we can choose to undertake it with the resolve and tenacity necessary to ensure that our changes are implemented and accepted. As with any such fundamental change, results may vary. Therefore, we must keep our eye on the ball and modify our approaches as necessary to ensure success.

It would be remiss to not discuss the fact that cultural change is a long-term challenge, and the practitioner should not expect that any substantial change will be implemented and undertaken in anything shorter then several years of effort. While you may be fortunate in being positioned in an organization that is hungry for change and willing to embrace modifications, the vast majority of practitioners will experience a slow and lethargic organizational shift that requires constant vigilance to ensure that changes continue to move the organization forward.

Bear in mind that this approach has been used successfully in multiple organizations to undertake this Herculean task, and that there is light at the end of the tunnel. Take the time to fully understand the problems with your culture, and mentally embrace the necessary changes to be implemented. With this studied coherence, project management cultural change is a practical and achievable goal. Remember: stay the course; all will be well!

Program Management

Program Management is an area of significant challenges to most technical organizations. The Program Manager is responsible for the entire life cycle of a product family or program within the organization. This program will likely be made up of many different phases, including concept refinements, technology development, systems development and demonstration, and production and deployment.

These responsibility areas extend far beyond the typical responsibilities of a project manager. In fact, many of these responsibilities extend well beyond the responsibilities of an engineering department. Therefore, the Program Manager must have a long-term viewpoint, and recognize that his or her role is one of steering the development of families of products or processes. This requires an ability to forecast and predict the behaviors of the marketplace, with enough forethought to allow the subsequent development of various systems and products prior to a shift in the marketplace.

Additionally, the Program Manager must also deal with many departments that extend well beyond the engineering function. They may include finance, accounting, marketing, sales, production, packaging, shipping, quality, and many others.

It is clear that the program management function operates at a much higher level than that of project management. Although many of the skills are identical, a typical program will incorporate many projects under its umbrella.

Therefore, the Program Manager is a manager of project managers in general. This requires a different set of skills than that discussed in the project management domain, and also requires a different mindset in the execution of the program. These differences will be discussed in detail.

Programs and Projects

We must understand the distinction between programs and projects. Projects are typically relatively small in scale, with concrete objectives and finite timelines. Programs, on the other hand, might extend for many years, and their expected outcomes may be unclear at their outsets. Therefore, programs are much more nebulous and vague than are projects.

Figure 97: The Program Life Cycle

Typically a program will spawn many projects in order to complete its diverse objectives. Many of these projects will reside within the Engineering domain, and many will reside outside. It is the role of the Program Manager to integrate these very disparate projects and perspectives in order to achieve a harmonious whole.

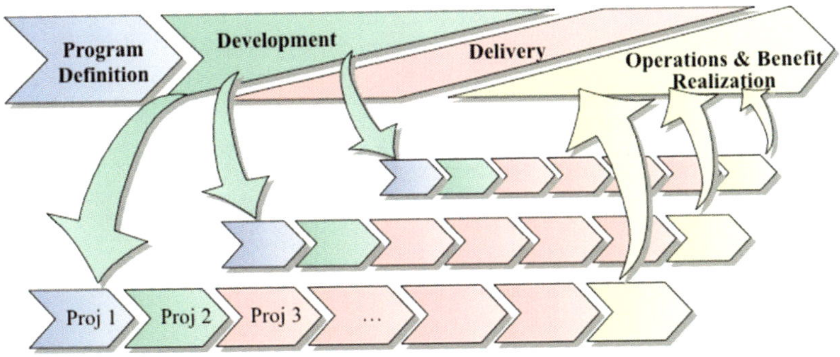

Figure 98: Programs spawn many Projects

Programs and projects, then, have very different life cycles and durations. Although the skills required to accomplish them are very similar, the outlook and perspective required of each is very different. The Program Manager is responsible for the long-term viability of the product and organization, while the project manager is responsible for the short-term completion of specific objectives as shown in **Figure 98**.

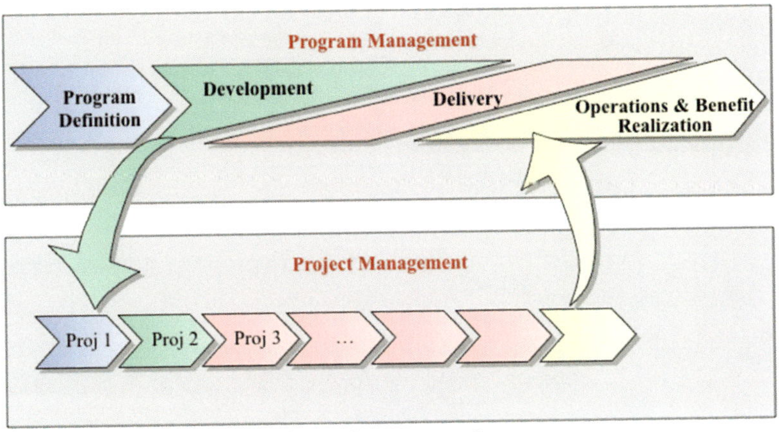

Figure 99: Program and Project Management

It becomes incumbent upon the engineering manager to accomplish engineering objectives in a practical manner, meaning that we will plan and execute our projects as previously discussed. In addition to these responsibilities, we must support the product roadmap, which will clarify the products that are on our development horizon, based upon the Program Manager's insights.

Due to the significant complexity of the program management function, the engineering leader must also recognize that his or her responsibility extends into any development that might alleviate the Project Manager's responsibility of overseeing projects (see

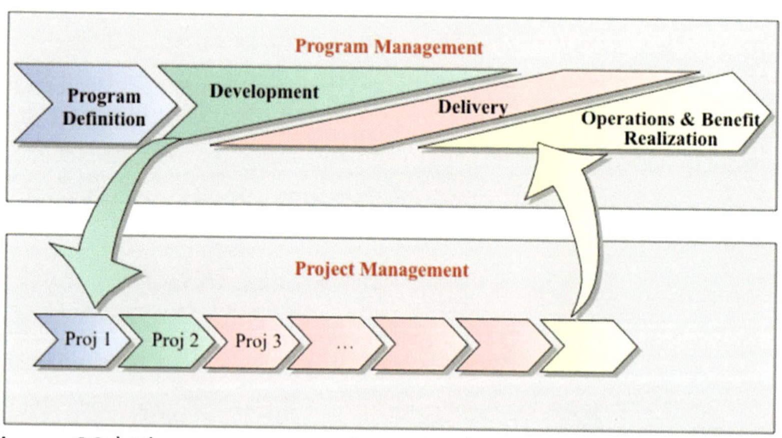

Figure 99.) Thus, a systematic and robust reporting system is necessary at the project level to support the program's objectives. This is another compelling reason why projects must be planned in detail and a realistic status must be defined and reported utilizing a business rhythm, as seen in **Figure 100**.

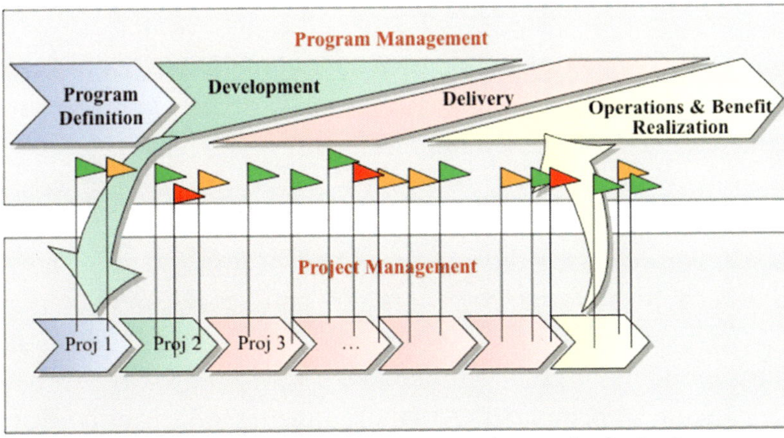

Figure 100: Project Status is Reported Regularly in Support of Programs

Program Management differs from Project Management primarily in scope and scale. There is a greater level of responsibility for the Program Manager, which requires a high level of business knowledge and capability. The magnitude of the differences between project managers and Program Managers is quite large, as the Program Managers must be very astute in political and diplomatic skill areas.

Program Management is clearly a step up from the Project Management domain. It is typically of more importance to the organization, deals with more complex efforts, and commands a higher degree of respect within the organizational hierarchy. Programs are strategic and typically include two or more interdependent projects that share a common and significant goal. The major goal of a program differs from a project, as it is focused on strategic alignment and business success. Therefore, the focus of the program is about competitive advantage and sales. Add to that fact that Programs include multiple projects; it is by definition of a higher level of complexity. Program Managers also are required to perform functions that go well beyond those of a typical project manager, such as defining the acquisition process and establishing the resources, funds management, contracting and logistics, to name a few.

The scope of the program may cover many different types of projects from many different divisions or organizations, each of which in some manner share the same goal. For example, a large company may develop a program to develop alternative energy methods in order to meet future worldwide energy needs. The program could include solar, wind, nuclear, biofuel, smart grid and other projects all directed towards this common goal. The magnitude of these projects and the program overall can be quite large. Each project within the program will retain its own goals, specification, schedule, and budget. It is clear the alternative energy program is designed to act as an umbrella to coordinate the efforts of all of these related energy projects within the program, providing benefits that may not be readily obvious if they are managed individually.

Programs are by definition larger than typical projects within an organization and therefore command larger financial resources and staff. If a typical project in an organization were made up of ten individual contributors at a cost of five hundred thousand dollars, one would expect the program in that same organization to cost perhaps several million dollars and constitute a hundred staff members. Very large programs have been known to grow to the size of several billion dollars.

Program Managers must bring skills in the areas of leadership and process development to the program. The leadership skills are similar to those of project managers when scaled up to the program level. In Program Management, leadership skills focus on several specific areas of responsibility:

- The development of a program vision that will be used to steer and engage project teams and their efforts
- The development of a program culture that establishes a trustworthy and professional work environment
- Motivation of the program contributors to reach a well-understood goal
- Collaboration across organizational boundaries in order to share resources

The key challenge for any Program Manager is to establish a balance between the leadership of the program and the processes necessary for its success. The major focus and responsibility of the Program Manager is the horizontal integration of all related efforts associated with the program, even though the projects within the program can be quite different in scale, scope and size, with appropriate integrated processes that are consistently applied across the organization. These activities vary by organization but include planning methodologies, communication plans, metrics and their generation, resource allocations and risk management methods.

Establishing program goals and priorities are major drivers in the integration and decision-making of any program. All efforts should be aligned and directed towards meeting the needs of a specific business strategy. The Program Manager assumes full accountability for the success of the program and achievement of business goals, and will also likely share in financial success. Successful Program Managers will maintain a clear connection between organizational strategy and program execution.

A major challenge for the Program Manager is to integrate and manage the processes within the program (and existing organization) to enable the achievement of program goals and objectives. A major portion of the Program Manager's time is therefore taken up in the creation and development of these processes. Coordination of the complex interdependencies of a program can be a big challenge, and yet management of all of these moving parts is necessary in order to achieve program goals. The demanding amount of coordination required in a complex environment is why the Program Manager is typically a highly skilled and capable individual.

A relatively simple structure can be used to describe the varied responsibilities that exist throughout the life of a typical program. Without a clear definition of roles and responsibilities, the vagueness of the requirements for each team member can result in a highly convoluted program effort. One technique used to describe these various responsibilities is a Swim Lane Diagram. An example of this is shown in **Figure 101**.

As can be seen in the diagram, the existing products are defined at the top level. The needed products and capabilities are defined at the Program Management level, which in turn defines when requirements must be established. The Systems Engineering effort must then be started, and so on throughout the typical project life cycle. In this example, a new product development program consists of approximately twenty individual projects spanning multiple functional areas.

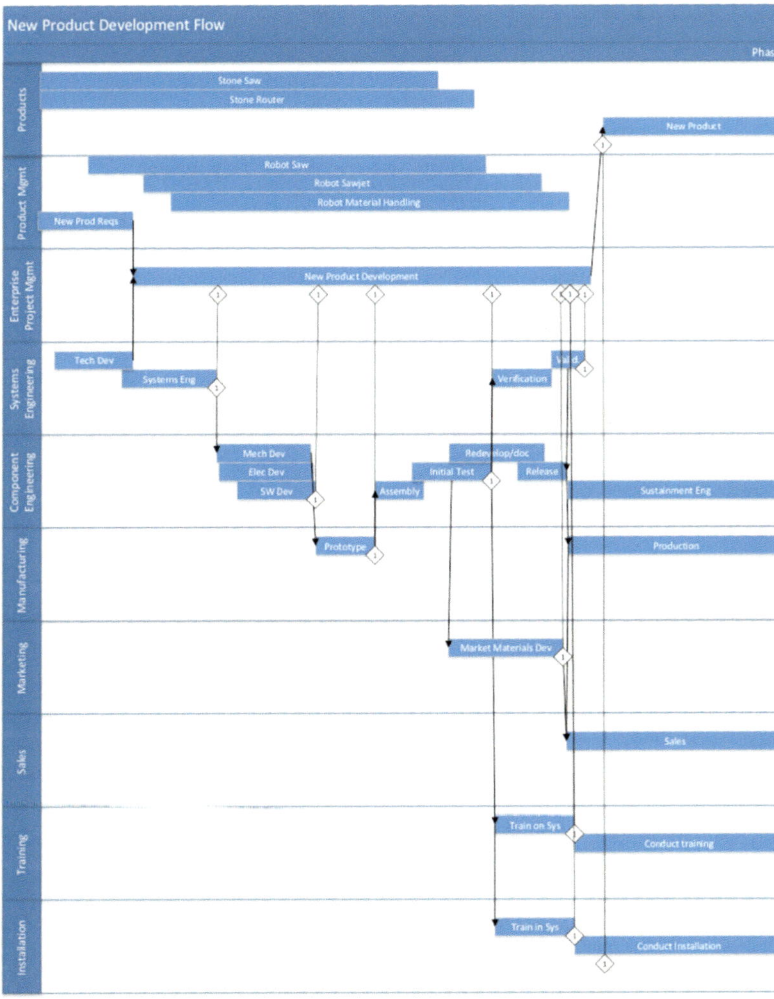

Figure 101: Program Management Swim Lanes

Program

A program is a group of projects managed in a coordinated way to obtain benefits not available when managing them individually.

Many programs will include elements of ongoing operations. For example: The ABC vehicle program includes both the project (or projects) that will design and develop the vehicle as well as the ongoing manufacturing and sustainment projects associated with that vehicle in the field.

Programs may also involve a series of repetitive or cyclical undertakings. For example, major construction companies often run cyclical construction programs, functioning as ongoing operations involving interrelated projects. Similarly, many nonprofit organizations run a fund-raising program on a regular basis to sustain their constant efforts to obtain financial support for the organization.

Project Portfolio

A portfolio of projects is a collection of projects owned or managed by the same organization that are only indirectly related. These projects may not be directly related with regard to inputs and outputs, meaning one project does not necessarily feed into another. Portfolios are typically chosen with a diversity of projects, some of which may be directly correlated and related while some may not appear to be related in any manner. Typically, organizations will construct a portfolio of projects utilizing an aggregated project plan made up of several different types of projects. The aggregated project plan will contain some projects that are focused on advanced research and development. Secondly, the organization may focus on new products or processes that result in breakthrough projects. Platform projects focus on advancements in product families within the organization. Derivatives focus on enhancements to existing products and processes.

Advanced R&D projects are typically identified as innovations in technology development that are necessary prior to embarking on a commercially viable product development. This is often referred to as pure research for the development of new sciences and methods.

Breakthrough projects are those that involve a significant change in the product and/or process to establish a new core product or process family that may result in several new products being introduced into the marketplace. These projects typically result in significant advances in market share or changes in market position for the business.

Platform projects focus on a known product family within the organization that will be utilized as a base for the design of new products or processes that are directly related to that platform. An example of this is shown in a typical automaker, where several different makes and models of cars are based on the same platform. This simplifies the basis of development of these complex systems.

Derivative projects are typically focused on cost reduction or product or platform enhancements for the development of more streamlined and cost-effective production processes. These projects are usually undertaken in incremental improvement efforts within the organization, and they are intended to keep the existing product line viable and profit generating for the near term.

Many resources exist that delineate the responsibilities of the Program Manager. One of the most complete of these can be found at the Defense Acquisition University website (www.dau.mil). This document, while being focused on acquisition programs for the government, nonetheless contains extremely valuable information for the Program Manager in any commercial organization.

The first area of responsibility for Program Management is the acquisition process and resources available. This means the project manager must understand the mechanism by which funding is allocated, how resources are applied, contracting methods, and the development life cycle of a project or program. For example, the Program Manager must be familiar with the five-phased approach to long-term program developments, including the concept development phase, technology development phase, product development phase, deployment and sustainment phase, and phase-out. Without a clear understanding of how the Program Management process is intended to function, a typical Program Manager will find it challenging to accomplish the program's objectives.

Life cycle logistics is also a crucial subject area, focusing on managing the program from cradle-to-grave, or through the entire life cycle of the program. This includes a great deal of planning, including the development of life cycle sustainment metrics, affordable system operational effectiveness, a product support strategy, integrated product development, and a model describing how product logistics will be accomplished within the business.

Sustainment issues are also typically managed within the program. These efforts are focused on planning for spiral developments and enhancements to the system based upon the typical development cycle of the technologies employed. For example, when an organization is producing a ship for the U.S. Navy that will typically last thirty to thirty-five years, it is clear that computing systems located within the ship have a typical life cycle of eighteen months and will need continuous refresh efforts associated with their constituent elements.

Program Management is also focused on the organization of the program office, employing varying structures. These include a traditional management structure, functional structure, pure product structure or a matrix orientation. In general, the most widely accepted construct in today's development environment is the use of integrated product teams or IPTs, which are a variation of a matrix organization.

The role of manufacturing in the acquisition process is typically defined very early, and is associated with defining the appropriate production methods and processes that will result in greatest benefits of the program. This also covers the domain of producibility, which considers the role of quality management systems and design for manufacturability in the early phases of the program's development.

Testing evaluation is defined throughout the program. This includes factors associated with modeling and simulation planning, and how those models and simulations establish the performance of the system in comparison with the expectations or needs of the stakeholders.

Program Managers are also fundamentally responsible for the appropriate planning, scheduling and control of the programs and projects under his or her umbrella. This includes many of the planning tools previously described, whereas Program Managers often must educate their various project teams to establish a standardized process and approach to planning and execution of projects. The standardized approach will enable the Program Manager to establish an appropriate business rhythm and reporting structure that will simplify the oversight requirements of the multitude of projects feeding into the program.

Risk Management is a key area for Program Managers, as they must establish the appropriate risk undertaken by the program such that they can maximize the return on investment for the organization. Included in this area are the definition of technical performance measurement metrics within the program, a Risk Management process, and various tools such as Root Cause Analysis, Fault Tree Analysis and Failure Mode Effects and Analysis. These can be used extensively to establish direction for the continued development of technologies on the program. The Risk Management process is a continuous process that will become the principal focus of the Program Manager during the execution of the program.

Systems Engineering is a large part of the Program Manager's domain responsibilities. This includes the systematic development of the users, user needs and system requirements prior to the system's design. Many other texts are dedicated to the domain of Systems Engineering, and these concepts are better discussed there.

Program Managers must also possess excellent leadership and managerial skills. These are necessary to properly organize and manage the program management office, as well as to delegate work amongst the resources provided. Program Managers must also be adept at coaching their direct and indirect reports and managing effectively within the integrated product team environment. By virtue of these needs, the Program Manager must clearly and concisely communicate the program's expectations, and how performance will be measured. Normally this is accomplished by means of a detailed team charter as well as generalized skills such as working effectively with groups, getting high performance from development teams, running effective meetings, employing effective communications, effectively managing time and utilizing continuous process improvement techniques.

Implications for the Leader

It is vitally important for the project leader to understand the roles and responsibilities of the various members of the program team. The leader must ensure that the Program Manager is accomplishing the tasks falling within the Program Management responsibilities, such as requiring that appropriate vision and strategy be provided to the project teams.

Secondarily, the project leaders must be sure that the projects executed within the engineering domain are supportive of the Program Management organization and its expectations. Therefore, although the project leader will require status updates of his or her project teams, they must be designed to allow for utilization by both the engineering department and the Program Manager. Clarifying these expectations and methods effectively for the team can greatly simplify the efforts and burden placed on the team by management and program. A description of appropriate roles and responsibilities for different stakeholders on a program or project is included for reference.

Roles and Responsibilities

The roles and responsibilities are defined for each of the typical roles on a project. These roles are then defined with regard to the typical tasks and responsibilities to be accomplished during each of the typical life cycle phases in a summary view. The specific role of the sponsor is clarified in further detail throughout the remainder of the document. These typical project roles include:

- Organizational Management
- Project Sponsor / Business Sponsor
- Program Manager
- Project Manager
- Project Team
- Customer

(from USDA Project Management Handbook, Version 1.0, online publication)

Organizational Management

Organizational Management is responsible for the identification of the need and opportunity for a project, assessment of project risk, and the approval of the project's feasibility and resources. It must also establish the strategic plans and validate that projects are consistent with customer and organizational requirements. Management provides close oversight for high risk or high cost projects.

MANAGEMENT ROLES AND RESPONSIBILITIES					
General	Initiating	Planning	Executing	Monitoring/ Controlling	Closing
Provide leadership and resources to establish and improve project management. Ensure that sufficient resources are available to conduct projects. Review and approve commitments to external entities (e.g., customers, vendors.) Ensure staff is properly trained in project management techniques and principles.	Select Project manager and assist in Project Team staffing. Review, validate and approve project charter. Authorize and provide funding.	Verify that project goals and objectives are defined. Review and approve Project Management Plan, cost, and risk, and establish management reserves. Provide management oversight as predicated by review of the project risk analysis, risk response planning and Project Management Plan. Enable project staff availability.	Regularly conduct executive management reviews and provide oversight.	Review project status and corrective action plans (if required.) Review and approve changes affecting scope, timing, cost, and/or quality, as required.	Validate project completion (goals & objectives.) Verify customer and sponsor acceptance. Review and close project accounting/ financial files. Review project lessons learned and post project reports for continuous improvement action.

Project Sponsor / Business Sponsor

The Project Sponsor is usually a member of the management team who will be the recipient of the project's end result (the product.) The Project Sponsor is typically the head of a program area. This individual makes the business argument for the project to exist, controls the overall funding of the project and defines the acceptance criteria of the product.

SPONSOR ROLES AND RESPONSIBILITIES					
General	Initiating	Planning	Executing	Monitoring/ Controlling	Closing
Articulate project and/or customer requirements. Validate that project requirements are met. Provide the necessary funding and resources as appropriate. Champion the project to provide exposure and buy-in. Communicate the sponsor's views on project progress and success factors to the Project Team and other stakeholders. Define project goals, objectives and success criteria.	Provide the strategic goals and objectives of the recipient organization and guidance to the Project Team to identify the relevance and value of the project. Obtain or provide funding for the project.	Review and approve the Project Management Plan and management approach. Participate in planning sessions.	Attend executive requirement reviews. Resolve escalated project require-ments issues; remove barriers and obstacles to the project. Provide written agreement to project requirement s and qualifying criteria.	Attend and participate as needed at Project Status Reviews and steering meetings. Attend change control meetings and review and approve changes in scope, timing, quality and/or cost as impacted.	Provide representation or input to lessons learned reviews. Sign off on project completion.

Program Manager

The terms “program” and “project” management are often used interchangeably. However, within this document the two terms and concepts are separate and distinct. Program Management is defined as a group or series of related projects and ongoing systems/applications managed in a coordinated way to achieve resource, cost and quality efficiencies not available to individual projects. Programs generally support strategic goals and objectives, while projects may be more targeted in focus. The Program Manager is responsible for the management of a series of related projects and the management of the corresponding project managers.

PROGRAM MANAGER ROLES AND RESPONSIBILITIES					
General	Initiating	Planning	Executing	Monitoring/ Controlling	Closing
Plan, organize, staff, direct, control and coordinate. Recommend composition of own program team. Own and guide the program Reward and recognize performance. Be accountable for cost, schedule, quality and scope. Resolve any outstanding issues among the Project Teams that cannot be resolved within the team. Responsibility for overall resource allocation for Project managers assigned to the program. Maintain ongoing communication with the Project managers from	Assign Project manager and assist in Project Team staffing. Review/validate/concur on project charter. Validate and communicate individual project objectives.	Verify that project goals and objectives are defined. Verify that project is aligned with the strategic goals of the program. Review/ concur with Project Management. Plan, cost, risk and establish management reserves. Provide guidance on cost and schedule development. Ensure project staff availability. Conduct routine program planning sessions as defined by individual's organization.	Conduct regularly scheduled project reviews.	Review project status and corrective action plans (if required.) Review, concur and participate in milestone review briefings. Review/ concur on changes affecting scope, timing, cost, and/or quality, as required. Prioritize any changes to project scope.	Review whether stated improvements or benefits were realized from the project. Assure customer and sponsor acceptance is obtained. Review and concur on project accounting/ financial file closeout documents. Review project lessons learned and post project reports for continuous improvement action.

the program management level perspective. Communicate project status to fellow Program Managers. Ensure applications are developed consistently with the project management cycle. Ensure projects are managed in accordance with the recommend-ations for project management as outlined in the Project Management Handbook. Ensure that security certification and accreditation (C&A) requirements are met.					

Project Manager

The project manager has overall project responsibility. In order to achieve success, he or she should work closely with the sponsor with respect to staffing requirements and funding availability. The project manager is also responsible for completing the project on time and within budget, and for meeting the quality criteria and requirements.

The project manager should be assigned as early as possible in the life cycle of the project in order to establish project ownership and management responsibility, as well as to begin the development of the project requirements from the "ground up."

PROJECT MANAGER ROLES AND RESPONSIBILITIES					
General	Initiating	Planning	Executing	Monitoring/Controlling	Closing
Comprehend and implement organizational project policies and procedures. Maintain project staff's technical proficiency and productivity, and provide training where required. Establish and maintain project quality. Identify, purchase and acquire project infrastructure needs. Identify and secure Project Team resources. Serve as focal point for project communi-cations. Ensure that security requirements are met.	Develop project concept document. Define the needs of the sponsor and of the organization. Develop Project Charter and obtain approval. Identify and document project constraints. Identify and document project assumptions. Document requirements. Review customer requirements for feasibility and available resources.	Develop Project Plan, including Project Charter, Scope Statement, constraints, assumptions, WBS defining project deliverables, cost estimates and project budget, major milestones, schedule, resource require-ments, acquisition/ procurement plans, risk analysis and response plans, Project Team structure and communications plan. Also included will be the deliverables acceptance criteria (quality metrics) and the	Manage and monitor day-to-day activity and provide direction to team members and supporting organiz-ations. Manage to and monitor quality targets and goals (both project and product.) Manage and monitor risk response strategies. Dis-seminate project inform-ation and maintain commun-ication. Develop and update system security	Develop and distribute project performance reports. Regularly review project status, evaluating performance criteria (scope, cost, schedule & quality.) Develop and manage corrective action plans Evaluate project performance and initiate change requests as required (scope, cost, schedule or quality.) Participate in change control board to review and approve product/ project changes. Review project risks and establish	Obtain customer and management approval and acceptance of completed product. Complete contract closure process. Close out open action items. Develop post-implement-ation report. Conduct lessons learned session and develop recommend-ations for continuous improvement. Close out any financial accounts or charge codes. Archive all project data. Recognize Project Team and celebrate success.

		acceptance process. Develop the supporting plans such as scope, cost, risk, schedule, quality, resource, security deliverables, procurement and change management plans. Obtain stakeholder approval and acceptance of the Project Management Plan. Obtain organization-al commitment and support for completion of project task assignments, timing and quality. Establish baselines.	plan and other security deliver-ables. Manage, or support purchases and acquisit-ions process and contract admini-stration requireme nts.	risk response plans. Adjust Project Management Planning, as required, to include approved changes in scope, timing, cost or quality after obtaining customer approval.	

Project Team

The project team is responsible for performing the project's activities. Project team members, as required, may assist the project manager with planning the project (for example, scope and WBS) and may also assist with obtaining commitments to complete the project within established schedule and budget constraints. Customers and/or stakeholders should interact with the project team to ensure requirements are properly understood and implemented.

PROJECT TEAM ROLES AND RESPONSIBILITIES					
General	Initiating	Planning	Executing	Monitoring/ Controlling	Closing
Identify product alternatives. Complete the project within budgeted cost, schedule and quality require-ments. Support Project Manage-ment Planning and control. Participate in identi-fying, miti-gating, and monitoring project risks.	Provide estimates for product deliver-ables. Analyze require-ments for clarity, complete-ness and con-sistency.	Develop technical approach. Participate in the develop-ment of the Project Manage-ment Plan. Participate in Scope Planning. Identify tools needed for project Identify staff training needs.	Create product and process solutions. Conduct or participate in internal and external reviews. Provide quality assurance support. Manage work effort to maintain on time, on cost and on quality results. Identify any project roadblocks, barriers or unanticipated risk events	Track the project execution effort and submit status reports (scope, cost, and schedule.) Maintain project and product quality require-ments. Identify and react to risk events as they are identified or occur. Participate in change control reviews.	Participate in lessons learned sessions. Identify ways to improve project processes or products (continuous improvement.) Turn over all project related documentation to the Project manager for archiving.

Customer

A customer is responsible for communicating project needs and verifying that requirements have been met at project completion. A customer may be internal, external or both.

CUSTOMER ROLES AND RESPONSIBILITIES					
General	Initiating	Planning	Executing	Monitoring/ Controlling	Closing
Articulate customer require-ments- Validate that project require-ments are met. Support and conduct staff training programs as required to ensure the staff is "ready to accept" the new product. Be pro-ponent of the new product to the customer.	Clearly define customer needs and require-ments to the Project manager and Project Team.	Review and approve Project Manage-ment Plan. Attend and participate in project require-ment reviews. Assign customer personnel as project points of contact. Provide written agree-ment on require-ments and qualifying criteria Provide input into deploy-ment planning.	Review project status reports Provide project support infra-structure as required.	Participate with Project Team developing corrective action plans addressing variances in time, cost or quality. Communicate identified issues associated with project or product performance. Validate quality assurance of deliverables. Participate in change control process. Review and approve or escalate project changes affecting scope, time, cost or quality.	Accept the completed product. Provide represent-ation or input to lessons learned reviews.

References

Anonymous (1994) Principle Reasons for DoD Software Project Failures, Report of the Defense Science Board Task Force on Acquiring Defense Software Commercially (1994)

Babcock, D., (1996) Managing Engineering and Technology, 2nd Ed, Prentice Hall, New Jersey

Badiru, A., (1993) Managing Industrial Development Projects, Van Nostrand Reinhold, New York

Bak, P. (1996) How Nature Works. Springer-Verlag. New York.

Belassi, W., Tukel, O. (1996) A new framework for determining critical success/failure factors in projects, International Journal of Project Management Vol. 14, No. 3, pp. 141-151

Covey, S., (1989) The 7 Habits of Highly Effective People, Simon and Schuster, New York, New York

Faulkner, S. (2006) Information System Engineering, Chapter 2, Feasibility Design. Catholic University of Louvain.

Hayes, R., Wheelwright, S., (1984) Restoring Our Competitive Edge, John Wiley and Sons, New York

Jones, C. (1995), Patterns of Software System Failure and Success, Intl Thomson Computer Pr (Sd) (December 1995)

Kerzner, H., (1995) Project Management, 5th Ed, Van Nostrand Reinhold, New York

Penner, D.,(1994) The Project manager's Survival Guide, Battelle Press, Columbus, OH

Peters, T. (1987) Thriving on Chaos. Harper Collins. New York.

Project Management Institute (2015) A Guide to the Project Management Body of Knowledge, PMI, Upper Darby, PA, 1996

Quinn, R. E. and J. Rohrbaugh. (1983). A Spatial Model of Effectiveness Criteria: Towards a Competing Values Approach to Organizational Analysis. Management Science , 29(3), 363-377.

Quinn, R. E. and K. S. Cameron. (1983). Organizational Life Cycles and Shifting Criteria of Effectiveness: Some Preliminary Evidence. Management Science , 29(1), 33-51.

Robbins, (1999) Essentials of Human Behavior. Houghton Mifflin, New York.

Rosenau, M., (1992) Successful Project Management, 2nd Ed., Van Nostrand Reinhold, New York

Shainus, Dekom, McVinney, (1994) Engineering Management: People and Projects, Battelle Press, Columbus, OH

Shtub, Bard, Globerson, (1994) Project Management: Engineering, Technology, and Implementation, Prentice Hall, New Jersey

Smyth, H., (1995) The Ten Natural Laws of Time and Life Management, Franklin Press, New York

Steudel, H., Desruelle, P., (1992) Manufacturing in the Nineties, Van Nostrand Reinhold, New York

The Standish Group (1996), Chaos Report, Boston Mass., www.thestandishgroup.com

Turtle, Q., (1994) Implementing Concurrent Project Management, Prentice Hall, New Jersey

USDA (2006) USDA Project Management Handbook, Version 1.0, online publication

Website (2017) Software Systems Failure Modes, http://suraj.lums.edu.pk, March 2017

Website (2017) The Chaos Hypertextbook, http://Hypertextbook.com/chaos, March 2017

Website (2008) Wikipedia. Requirements Traceability. June 2008

Website (2009) Personality Types Equivalents. http://www.ylcf.org/you/personality-equivalents.htm, June 2009

Website (2009) Personality Types. http://macdl.org/edu/socmgmnt.htm, June 2009

Website (2017) Maslow's Hierarchy of Needs. Wikipedia. March 2017

Wheatley, Margaret (2006) Leadership and the New Science: Discovering Order in a Chaotic World, Berrett-Koehler Publishers; 3rd edition, New York, NY

Zimmerman, F. (1997) University of St. Thomas. Personal Communication. Saint Paul, Minnesota

CPSIA information can be obtained
at www.ICGtesting.com
Printed in the USA
BVIC01n0754071217
501952BV00013B/12

* 9 7 8 0 9 9 8 1 4 4 2 4 5 *